# Hotels:
# Between the Lines

时尚艺术酒店

(澳)斯考特·惠特克 编　　于芳　李红　译

辽宁科学技术出版社

Scott Whittaker
group executive director, creative director &
founding partner
email: scott.w@dwp.com

### Key Experience

Educated in Australia with a Bachelor's degree in Architecture, Scott moved to Asia in 1994, and, together with his business partners, transformed dwp from a local Thai-based company into an award-winning, pan-Asian design firm, reaching out further still to the Middle East and Australia. Innovation, creativity and energy are the cornerstones of Scott's success. Within Thailand, he has led a team that has amassed many accolades for diverse projects across sectors. Some notable highlights include such renowned hospitality projects as the highly acclaimed Sirocco Restaurant & Skybar (featured in The Hangover Part II movie), Mezzaluna, State Tower and 87 Plus at the Conrad, Bangkok, as well as the highly acclaimed Earth Spa at Evason Hideaway in Hua Hin, Thailand. Now, at the helm of all group creative activity, Scott oversees and contributes to the success of projects across industries and disciplines, injecting enthusiasm into every single venture. Scott is skillfully able to translate clients' visions into cutting-edge designs, whilst simultaneously maintaining and strengthening the creative vision, design standards and design direction for the group.

斯考特·惠特克

斯考特·惠特克是全球设计合作集团(dwp)的全球设计总监、执行总监兼创立者。在澳大利亚获得建筑学士学位后，斯考特在1994年移民亚洲，与他的合作伙伴们一起让dwp这家泰国的设计集团变成了著名亚太地区的获奖设计集团，并将事业开展至中东和澳大利亚。创新、创作力和热情是斯考特成功的基石。在泰国，他设计的不同类型的项目受到了广泛的赞誉。其中参与过的著名项目包括备受赞誉的热风餐厅和天空酒吧（曾出现在电影《宿醉2》中）、月怡餐厅、洲际塔莲花餐厅以及在曼谷康拉德酒店的87加酒吧，同样还包括广受好评的泰国华新酒店的艾梵森水疗中心。现在，斯考特掌管着集团所有的创造性活动，监管着各领域中的项目创意，为它们的成功做出贡献，并为每一次探索注入激情。斯考特能将客户的想象转化成先锋设计作品，与此同时，集团的创意理念、设计标准和设计方向被其完善和强化。

# CONTENTS
目录

006 / Branding & Longevity of the Design Hotel Experience
设计酒店中的品牌化与长效性研究

008 / Anticipating Desire
预想的渴求

010 / Introduction
导论

## Hotels in Fancy Dress
穿奇装异服的酒店

016 / Endless Revival of Old Things, Fashion Concept in Antique Building
让旧事物无休止的复兴，古董建筑中重新孕育的时尚思维
Hotel Le Bellechasse
柏歇斯酒店

028 / A Colourful Time Created by the Palette Master
调色大师的彩色时光
Hotel du Petit Moulin
小磨坊酒店

038 / The Trompe l'Oeil by Art and Colour, The Birth of Neo-Braoque
艺术与色彩的视觉陷阱，新巴洛克的诞生
The Hotel Le Notre Dame
圣母院酒店

050 / Surrealistic Sweet Dream
超现实美梦主义
Maison Moschino
莫斯奇诺酒店

064 / "Unfinished" Aesthetic Values from the Genius of Deconstructivism
解构怪才的"未完成"美学
La Maison Champs Elysées
香榭丽舍家园酒店

074 / A Story of Enchanted Hotel Comes True
酒店故事，魔幻成真
Hotel ORiginal
原创酒店

## Fashion Brands Hotels
时尚品牌的酒店业入侵

084 / The Hotel Design Dream of the Giant in Fashion Industry
时装界巨子的酒店设计之梦
Armani Hotel Dubai
迪拜阿玛尼酒店

094 / The Brilliant Rebirth of Colours and Patterns
色彩与图案的舞动新生
Hotel Missoni Edinburgh
爱丁堡米索尼酒店

104 / Medusa's Palace
美杜莎的神殿
Palazzo Versace
范思哲宫殿酒店

112 / Original Fashion
独创时尚
Casa Camper Barcelona
巴塞罗那营地之家酒店

124 / Fairy Shoes Home
仙履家园
Hotel Lungarno
朗伽诺酒店

132 / The Portrait of Ferragamo
菲拉格慕的自画像
Portrait Suites
肖像套房酒店

140 / The Eclecticism in Cartagena
芭蕾舞者的摇滚天堂
Tcherassi Hotel + Spa
切拉西酒店及水疗中心

## Timeless Classic of Fashion
时尚演绎的永恒经典

152 / A Gem of Art Deco Wearing Wrap Dress
穿上裹身裙的装饰艺术瑰宝
Claridge's
克拉里奇酒店

162 / **The Renovated Classic by a Fashion Generalist**
时尚多面手的创新经典
Alma Schlosshotel im Grunewald
阿尔玛格吕内瓦尔德皇宫酒店

174 / **The Fashion Collection of Dorchester**
多尔切斯特的时尚精选
The Beverly Hills Hotel and Bungalows
贝弗利山庄酒店及别墅

## Hot Destinations of Celebrities
名人们的时尚聚集地

186 / **Paris Is in the "WOW" Now!**
巴黎WOW时尚进行时!
W Paris – Opéra
W巴黎歌剧院酒店

196 / **The Fashion Growing in the Wild**
让时尚在自然中延续
UXUA Casa Hotel
UXUA之家酒店

210 / **Dominican Socialite Holiday Feeling**
多米尼加的名媛度假风情
Tortuga Bay
龟岛海湾酒店

220 / **The Artistic Practice in Architecture by Hat Magician**
帽子魔术师的空间艺术实践
The g Hotel
g酒店

230 / **The Eternal Legend in May Fair Street**
梅菲尔街上的永恒传奇
The May Fair Hotel, London
伦敦梅菲尔酒店

## Hotels: Painting Rooms
酒店内的绘画空间

240 / **A Micro World of European Arts**
欧洲艺术的微型世界
Hotel BLOOM!
繁盛酒店

250 / **Art Graffiti for Toronto**
多伦多的艺术风情涂鸦
Gladstone Hotel
格拉德斯通酒店

260 / **The Sculpture Graven by Scar**
伤疤雕刻的艺术品
Casa do Conto, arts&residence
故事酒店，艺术与住宅

## Hotels as Art Works
酒店艺术品

272 / **Salute John Olsen, Salute Australian Art**
向约翰·奥尔森致敬，向澳洲艺术致敬
The Olsen Hotel
奥尔森酒店

282 / **English Art Hotel in New York**
在纽约的英式艺术酒店
The Crosby Street Hotel
克罗斯比街酒店

290 / **Mobile Art**
移动的艺术
citizenM Hotel Bankside London, UK
伦敦河岸区居民M酒店

300 / **A Journey to Diverse Arts Neo-Baroque**
新巴洛克的多元艺术之旅
Byblos Art Hotel Villa Amistà
比布鲁斯艺术别墅酒店

310 / **European Art Deco Gem with Oriental Feelings**
充满东方风情的欧洲装饰艺术瑰宝
Prague Art Deco Imperial Hotel
布拉格皇家装饰艺术酒店

318 / **Index**
索引

# Branding & Longevity of the Design Hotel Experience

设计酒店中的品牌化与长效性研究

The design boutique hotel concept has followed a distinct and marked trend in the hospitality industry. Designers are putting their unique stamp and brand on luxury boutique hotels and opulently oriented hospitality experiences. It could be said that the boutique hotel concept started as far back as with Ian Schrager, one of the owners of the iconic nightclub 'Studio 54', and the founder of the Morgans Hotel Group, Ian Schrager Hotels and the newest Ian Schrager Company. His first hotel, 'Morgans', opened in 1984, marking the birth of the boutique lifestyle hotel, in collaboration with Andrée Putman, the renowned late French interior and product designer.

Later, further collaborations between hotel developers and the famous French product designer, Philippe Starke, radically popularized the trend, which took the elements of stage and club lifestyle and transformed it within a hotel or resort setting. The hotel became a hip, stylish and fun place to be, see and be seen in. Since then, the trend has expanded beyond the realms of the product designer to the fashion labels of such brands as Bulgari, Armani and Missoni, to namedrop but a few.

The possibilities are endless, given the wide variety of product and fashion labels in existence and the outcomes are as varied as the brands that may move into the hotel arena, from pop culture to haute couture, hipster chic to lavish luxury and enviable brands that represent wellness and sustainability. High-end design firms are often engaged to assist in realizing correspondingly befitting architectural and interior design environments, to match and ensure the integrity of the exacting standards of the brand in question.

The trend has since spread among all hotel types, with many of the global chains developing a designer brand of their own. Designer hotels were also typically smaller entities, but have since expanded to 400-bed accommodation, in chains, such as Starwood's W Hotels. The trends, however, are influencing all hotels from the budget to the luxury range, working with a household brand, instantly recognisable designer name. Subsequently the entire lifestyle of brands has moved into hotels, with names such as the likes of Versace, across the globe.

The hospitality industry is not the only sphere affected, as the tendency is also reverberating through to the serviced apartment and branded residence concept. dwp | design worldwide partnership (www.dwp.com) is working with globally iconic fashion brands, such as Ralph Lauren, Hard Rock and FTV, developing architecture and interiors for designer hotels and serviced residence developers.

As celebrities dominate our news and media today, designer brand hotels and the experience offered allow guests to themselves feel and be treated like a celebrity, star or supermodel for the duration of their stay. Hotels are a platform for living out both actual life and heady aspirations. Guests' use of hotels can be as much an expression and extension of their personalities, as an escape from regular life.

The key, therefore, to creating design hotel concepts and interiors is in understanding the lifestyle of the guests, the aspirations of the intended guest profile and the core values of the designer brand. Beyond this, designer hotels need to fulfill all the needs of a regular hotel, such as guest comfort, services, functionality and fire/life safety, etc. Great opportunities also exist for high revenue outlets, such as café, restaurants, clubs and bars, which need to be explored in depth. In short, the designer has to ensure theoretical, aesthetic design translates to practical and successful environments.

Fashion changes seasonally, so the hotel needs to represent the core brand values and lifestyle, rather than the current season trends, since hotels tend to have lives of 10 to 15 years, before renovation becomes necessary. While incorporating the essence of a brand's collection, the transposition of the brand ideals into a physical space, with rooms, restaurants and spas, is key to the hotel's success and the essential and correct lasting brand representation. The ultimate expression must be in the lobby/reception space, as it becomes the living room for the hotel and sets the scene for the overall atmosphere.

Wow-factor is important, but creating the right atmosphere is critical, right down to the service, staff and details, such as music, uniforms, graphics, lighting, menus and bathroom amenities. Beyond pure aesthetics, the design hotel guests must feel comfortable and reassured of the practicality and functionality of the environment, if the guests are to return.

This trend is not merely a passing fad, but has become a major influence on all hotels. That being said, the designs are rapidly evolving, and some designer hotels remain iconic and relevant, while others, focused too intently on current design crazes, quickly fade. If the designs remain relevant to lifestyles, with focus on the creation of aspirational layouts, architecture and interior spaces, as well as adopt an intensely end-user-oriented approach to the design solutions, they are likely to endure.

Scott Whittaker
Group Design Director, Executive Director & Founding Partner,
dwp | design worldwide partnership
(www.dwp.com)

在酒店设计业，设计精品酒店的概念已经成为了一种独特、明显的潮流。设计师们将他们独一无二的设计标签、设计品牌与奢华的精品酒店紧密相连，极大的丰富了酒店设计的体验性。精品酒店的概念起源可以追溯到伊恩·施拉格的身上，他是地标夜总会"录音室54号"的业主之一，也是摩根酒店集团、伊恩·施拉格酒店和新成立的伊恩·施拉格公司的创立者。他的第一家酒店"摩根"，与法国著名的室内和产品设计师安德利·普特曼合作设计，于1984年开业，这意味着第一家精品酒店的诞生。

不久之后，更多的此类合作在酒店业主和法国著名的产品设计师之间诞生，菲利普·斯塔克让这种潮流迅速的流行起来，他将舞台和俱乐部的设计模式引入到酒店和度假村的设计中。这类酒店成为了一种时尚、流行的娱乐及交友场所。从这时起，这种潮流从产品设计蔓延到时尚品牌的设计，例如宝格丽、阿玛尼和米索尼，也包括一些名气小些的时尚品牌。

如果设计产品和时尚品牌本身具有多样性，从流行文化到高端时尚，从潮流时尚到无尽奢华与令人艳羡，如果这些代表着健康和持续性的时尚品牌能让更多的设计成果加入到酒店设计的竞技场中，那么可能性就是无限的。高端设计公司通常提倡相应地采用适于建筑和室内环境的设计策略，再与品牌严格的标准理念相搭配，相融合。

当这种潮流影响到各类酒店的设计时，许多全球联营企业都发展了自己的品牌酒店。设计酒店虽然是典型的小规模酒店，但发展至今也有例如喜达屋W品牌酒店这样多达400个房间的酒店。这种潮流广泛的影响着所有酒店的设计，从预算到奢华等级，更有与知名品牌，甚至是明星设计师合作的酒店。随后，明星设计师的整个品牌理念都被融入到酒店中来，酒店也被冠以品牌的名称，例如享誉全球的范思哲品牌酒店。

受到这种潮流影响的并非只有酒店设计业，酒店式公寓和品牌公寓的设计理念也受到了影响。dwp设计事务所就正与全球的顶级时尚品牌合作，例如拉夫·劳伦，硬摇滚酒店集团和时尚电视台，为这些品牌酒店和酒店式公寓作建筑和室内设计。

当名人们控制着我们当今的新闻和媒体，时尚品牌酒店和它所提供的体验能让客人在住宿时感受到自己也受到名人、明星或者超模般的待遇。酒店是一个可以同时展现真实生活和实现幻想的平台。住在一家酒店，客人们最想的是能够尽兴的游玩、尽情的释放，酒店成了在客人日常生活之外的天堂。

因此，创作这类设计酒店的理念和室内设计的关键是去理解客人的生活方式，去了解目标客人的愿望和时尚品牌的核心价值理念。除此之外，时尚品牌酒店应具有一家普通酒店所有的全部基本元素，例如舒适感、服务、功能性和防火功能或安全性等。高端的消费场所，例如咖啡厅、餐馆、俱乐部和酒吧，这些是需要设计师深入挖掘设计的地方。简单的说，设计师要确保将理论、美学设计理念转变成切实又成功的空间场景。

时尚瞬息万变，因此这类酒店需要代表品牌的核心价值和生活方式，而不是品牌某一季的时尚潮流，因为酒店的一种设计潮流会延续10至15年，在潮流退去之时，翻修就变得必要了。当在酒店的设计中融入一个品牌的精髓，将品牌的理念移植到实体空间、房间、餐厅和水疗中心区，就成为了一个酒店成功的关键，也是延伸品牌价值不可或缺的关键。最能表达出品牌价值的功能区是大堂和接待区，这里作为酒店的起居室决定着酒店整体的氛围。

令人惊喜的元素是重要的，但是营造出恰当的氛围是关键的，直接与此相关的是服务、团队和细节，例如音乐、制服、图形、灯光、菜单和浴室设施。除去纯粹的美学观点，如果客人想要再次光临酒店，说明这类设计酒店会让客人感受到舒适，他们能在这里体会到令人安心的实用性和环境的功能性。

这种设计的趋势不会仅仅是一种时尚潮流，它会在整个酒店设计界产生影响。这正说明，设计界是在快速发展的，那些保持了标志性理念和相关理念价值的时尚酒店会得到成功，相反，那些只专注于潮流设计的酒店则会很快退出人们的视线。如果设计能够坚持关注人们的生活方式，关注布局、建筑和室内空间这几方面的创作，与此同时采用针对客户需要的设计策略，这样才能使这类酒店保持繁荣。

斯考特·惠特克
全球设计总监，执行总监兼创立者
全球设计合作集团（dwp）
(www.dwp.com)

# Anticipating Desire

预想的渴求

Since the days when the Pharaonic Kings covered their temple walls with pictographic images of themselves and their subjects, and Greek carvers created Caryatid columns idealizing the feminine physique and its beauty, mankind has created structures influenced by the human form and its fashions.

The allure of fashion has the ability to transform an individual into a character, to allow one's stamp of personal style and wit to create an aura of mystique, sensuality and connection. Through the influence of fashion, architecture relates to humankind on a more personal level.

Our culture has always been shaped by cross-pollination. The earliest known monuments and buildings were not attributed to either architects or designers - professions that did not exist until later. It was the rulers, the queens, the sultans, the kings, the patriarchs, and the priests who conceived the world's most notable ancient spaces. Astronomers, inventors, artists, politicians, philosophers and scientists directed the creation of buildings, influenced by an entire universe of scholarship, philosophy, and prevailing ideas of the time.

In today's highly connected and postmodern world, the cross-pollination of concepts is even more inevitable. Fashion influences architecture, architecture influences video games, video games influence movies, movies influence interior design, and interior design influences fashion. We revel in the works and wonders that cover our world, and in the fascinating beauty that we witness around us every day. It is commonplace for a well-known interior designer to create consumer packaged goods, or a street artist to become a lifestyle brand.

Despite this long history of the cultural influence of architecture and design, however, hospitality environments, namely hotels, have long been largely closed to influences from outside the industry. Hotels were always created by proprietors or families who owned them, with the help of highly specialized designers. Save for a select few mavericks like Conrad Hilton, new ideas were rare and risky and the hotels of yesteryear (with a few notable exceptions) were simply polished idealizations of local domestic settings.

It took a few idealistic individuals to break that initial mold.

The American Modernist architect John Portman believed that grand sculpture, art, and the sky itself should be integrated into the design of glorious hotel interior spaces. The venerable Frank Lloyd Wright drew from exotic foreign styles to handcraft a small number of beautifully intricate hotels replete with their own signature furniture, silverware, ceramics, furnishings, and fabrics. He designed everything in the space, in attempt to truly develop a lifestyle to fit his image of the way we should live, work and rest from head to toe.

While neither Portman nor Wright have their work depicted in this contemporary collection of projects, today's hotel designers owe thanks to these groundbreakers that came before them. With their precedent, they changed the business of building hotels, allowing for and provoking greater inspiration that defied convention in their day. They made it possible to not only think outside the box, but to eliminate the box entirely and work in the entire universe of ideas and experiences when conceiving spaces of leisure and escape. As the highly influential fields of now — design and fashion — increasingly find inspiration in one another, a selection of current works stand out where these influences collide.

The projects in this book represent a snapshot of Now. These featured hotels are the newest and the most immediate. Some are created for a smart business stay or even a night out with friends. Others are meant to be provocative — to create a distinctive impression — almost a mystique. Yet the world's best hotels are about detailing and craft, offering a balanced composition and refined palette. And just like the fashions that influenced them, some may have a distinct season, celebrated hotly and talked about by the young and beautiful, while others may live on for decades, canonized in the annals of media and memory. Each one is distinctive with its own perspective of a unique world composed from the mind and eye of its creator.

These are the environments that, for the time they are donned, transform the guest into someone else, becoming an aspirational image of his or herself. Like great fashion, these designs assume a theatrical quality that lifts us out of the everyday and creates distinct moments where our best qualities are magnified and become hyper-real.

When I approach a hotel design, I start with a narrative. I think, dream and plan for the guest. I imagine what they'll wear.

Then I look at the building. What is its history? How has its story evolved overtime? And how can we as designers pull the threads between creating a sense of place that feels organic in its culture, excites the traveler and leaves them feeling both cared for and inspired? Hotel design, at its best, is about predicting a set of wants and needs in the future. It is our job to anticipate desire and create a sense of identity – to draw a memorable experience for them.

I encourage you, the reader, to venture out and try one of these hotels for a day or two. Revel in who you become for that short time, even for just a moment.

Michael Suomi
Principal & VP Design
Stonehill & Taylor

埃及法老将埃及象形文字和当时的日常品刻画在庙宇的墙壁上以作装饰，希腊雕刻者创作了女像柱来展现完美的女性形体和容貌，可见，自从那时起人类就开始用人形和当时的时尚元素装饰建筑。

时尚的诱惑力是有能力让一个人变成一个人物，允许个人的风格和智慧绽放出光芒，使之充满神秘、感性的色彩，又显出关联感。受到时尚的影响，建筑在一个更个性的层次与人类更加紧密的相连。

我们的文化总是受到外来文化的促进，好比"异花授粉"。最早的知名纪念碑和建筑并不是因为建筑师或者专业设计师的功劳而被人铭记。统治者、皇后、苏丹、国王、主教和牧师，是他们构想出世界上最著名的古老空间。天文学家、发明家、艺术家、政治学家、哲人和科学家为建筑的创作把舵，令建筑受到整个文明、哲学和当时流行元素的影响。

在如今信息发达的后现代时代，"异花授粉"的理念更是不可避免。时尚影响着建筑，建筑影响着电子游戏，电子游戏影响着电影，电影影响着室内设计，最后室内设计影响着时尚。在设计师的作品和那些遍布在世界的奇迹中，在我们每天都会见证的美景中，我们都会体会到这一点。知名的室内设计师创作货品包装，或者一位街头艺术家创立了生活用品品牌都是习以为常的事。

先不顾建筑和设计受到文化影响的长久历史，单是酒店环境的设计，已经长时间的受到外界产业的影响。酒店通常是在专业设计师的帮助下，由业主或者产业家族主导设计。除了一些如康拉德·希尔顿的特例，新的理念在过去是稀有且充满风险的，过去的酒店（也有一些著名的例外）通常将当地的家装布置作为设计基础，再加以简单的润色"打磨"，使之理想化。

这种初始的设计模式由于几个理想的特例而被打破。

美国现代主义建筑师约翰·波特曼认为，大型雕塑、艺术和天空本身应该与夺目的酒店室内设计空间相融合。令人尊敬的弗兰克·劳埃德·赖特从异域风格吸取灵感，并亲手起草设计了几个别致的酒店，在其中布置并装点了他亲自设计的标志性家具、银器、瓷器、陈设品和织物。他设计了空间中的一切，尝试着真正的发展出一种与设计师本身形象吻合的生活方式，客人会追随着按照这种方式生活、工作以及全身心的休息。

无论是波特曼还是赖特，他们的作品都是用现代手法描绘，当今的酒店设计师也得益于这些创新者的突破。在这些开拓者的带动下，酒店建筑业发生了改变，当今的酒店建筑允许并鼓励更大的创新和对传统的挑战。这些特例使设计师不仅仅突破酒店设计的固定模式，更指引他们抛去整个固有的设计模式，能在整个设计的宇宙中自由地构想，设计出休闲胜地和世外桃源。目前最有影响力的领域——设计和时尚——正日益地在相互的身上找寻着灵感，一系列的优秀作品由于这种相互的影响而诞生。

本书收录的项目正是如今这种趋势的一个缩影。这些酒店是最新、最能体现这种趋势的。一些酒店是为了商业便餐或者朋友聚会而专门设计。其他一些酒店的设计目的几乎是一个迷，好像只为了创作出一种特别印象。世界上最好的酒店都是注重细节和工艺的，在架构上维持平衡并精心配色。这正如时尚，一些品牌会在某季大放光彩，受到年轻人和漂亮女士的欢迎和谈论，同时还有一些品牌拥有久盛不衰的口碑，受到媒体和时间记忆的推崇。每一种在他们各自的独特视角下都是与众不同的，这些视角正是创意者用思想和洞察力创作出的独一无二世界。

这些由设计师主导的酒店环境将客人转变成其他角色，让他们成为自己梦寐以求的那个人。这也如时尚一般，时尚设计拥有一种戏剧性的本质，它能让人们在日常的生活中超脱出来，制造出特别的时刻，激发出人们的最佳状态并超越自我。

当我接手一家酒店的设计工作，我会从描述一个故事开始。我为客人思考、梦想以及计划。我想象客人们的穿着，然后再观察建筑。它有着什么样的历史？它的故事是怎样被代代相传？对于我们设计师而言，怎样借以建筑中承载的文化创作出场所感，怎样让旅行者感到兴奋并带给他们关切和灵感，又怎样能将以上两种感觉相连？酒店设计，它的最佳状态应该是去预测未来的一系列所想以及所需。我们设计师的工作即是去预测一种渴求，创作一种身份感——为他们去描绘一场难忘的体验。

尊敬的读者，我鼓励您去探索，去尝试，到我们收录的任何一家酒店里体验一两天。即使仅仅一刻，请纵情在您的新角色里。

迈克尔·苏奥米
史通西尔与泰勒建筑公司主创设计师兼副主席

# Introduction

导论

Dressed up like a Persian beauty, the word "crossover", which leaves much to our imagination, is hard to define. Indeed, the word "crossover" catches our attention, impelling us to appreciate and study it in its brilliant presence in various fields. For a long period of time, multidisciplinary research has exerted a profound influence on the birth of a great number of new subjects. For example, in the field of literature, a crossover study of feminism and narratology laid the foundation of feminist narratology theory. Likewise, the word "crossover" also finds expression in the integration of different types of music, such as blues, the fruit of swing and jazz. In this book, we mainly discuss crossover in the design field.

It is because many designers' desires are much more directly embodied via a collage of diverse forms that many typical examples of crossover can be seen in the field of design. Another cause accounting for its popularity in this field is originality. Matthias, the German plan design master, once suggested that current originality tends to assemble seemingly irrelevant fragments together. Therefore he came to the verdict that "crossover is design, design is crossover". Nowadays, the complex design market requires that in many different situations a satisfactory result is hunted down by means of the use of many different fields of knowledge. In a word, the value of crossover is to inspire creative sparks and develop a sense of novelty by integrating separate fragments or even conflicting elements within a traditional design concept.

The birth of crossover design hotels offers ordinary tourists the access to enjoying their charm. The crossover design hotel stems from the wave of design hotels which forecast the trend of world hotel development. Since the day when design hotels emerge, people gradually come to realize that hotels are a place where people find inspiration and experience wonders. Such hotels put great emphasis on design, and by means of customized and individual interior design, they capture attention of distinguished guests. In the process of their growth, experimental economy is little by little developed. Guests accommodated in such hotels are eager to experience more than a bed or a meal. On the one hand, fashion enthusiasts expect a life surrounded by famous brands they are fascinated by. On the other hand, artists have a passion to receive sudden inspiration in the hotels they accommodate.

Besides, tourists who are in pursuit of high quality and novelty frequently visit such hotels. All these factors contribute to the birth of fashion and art hotels. To a certain extent, these crossover design hotels are an extension of design hotels and the carriers of art forms.

Corresponding to the rapid update in the fashion field, design is fast changing. As a result, it is reasonable to unite interior design, fashion and art. Current crossover design hotels are basically classified into two patterns: the fashion forward hotel and the art forward hotel. Hotels of the former pattern which gain popularity among prominent figures are the hottest. There is no lack of precedents in the crossover research of fashion and architecture. Initially, some fashion designers noticed some relevance between clothing and architectural design. Pierre Balmain, former aide to Christian Dior and the founder of the top-end brand Balmain, once remarked that "clothing is a kind of moving architecture". He was also the first clothing designer who was influenced by architectural structure. From then on, clothes in geometrical shapes with tough appearances could be found in some Balenciaga and Pierre Cardin designs, and new architectural elements were blended into clothing design, material and silhouette, which opened the gateway to the exploration in the fashion field. Correspondingly, some architectural designers also attempted to interpret their designs from the fashion perspective. For example, Herzog & de Meuron once made it clear that their architectural concept was inspired by material choosing in clothing design. Another example is the office buildings designed by Office dA in South Korea. Some sewing tips and techniques are employed in designing architectural structure and choosing architectural materials. Influenced by clothing design, such buildings made contributions to the architectural diversity, pictured the future development of hotel design and proved that it is of great possibility for fashion to set up a dialogue with architecture.

In recent years, the communication between fashion and architecture concentrated on the upsurge of the hotel interior design carried out by fashion designers. Fashion brands such as Armani, Versace, Missoni and Bvlgari are now starting to establish their fashion brand hotels, which supplement some commercial elements to the word "crossover". Relying on the former exploration in crossover research, such brands have the reason and ability to erect a fashion milestone on their road towards commercialization of crossover design.

In the fashion field, one of the most effective approaches to reinforce a brand in the consumer's mind is to experiment widely in the market. Usually, a clothing design brand, even slightly influential, involve in the design of perfume, bags and suitcases, shoes and furniture and housewares in the meanwhile. Besides, under the strong impact made by world-wide economic crisis, fashion brands feel impelled to enlarge their design fields and the crossover design hotel, for everything is arranged, naturally becomes their next target. Amazingly, how do the designers integrate fashion elements with interior design? The answer is compatibility and more precise, there exist some similarities in both fashion and interior designs. Todd Oldham, a versatile fashion designer, believes that on the subject of interior design, fashion designers are better tailored to create a more pleasant environment where guests feel like to be more fashionable.

The art forward hotel, the other pattern of current design hotels, which attracts attention from the public as well, can be generally subdivided into two types. The first type is designed by groups of artists invited and arranged by the hotel's manager. In many cases, the owners or the managers of this type of hotels are art lovers or artists themselves. Like Gladstone Hotel in Toronto, each guestroom of it is originally designed by one artist whose artistic inspiration cannot be duplicated. Similarly, the same pattern of hotel design was also adopted in the New Majestic Hotel in Singapore. As to another type, hotels are changed into art museums by their designers. In those hotels, works created by famous artists are placed everywhere, which makes guests feel artistic atmosphere and available to appreciate these works. Such hotels are warmly welcomed by artists. At present, there are special some art hotels only for artists to design. In China, Swatch Art Peace Hotel in Shanghai, Grace Beijing Hotel in Beijing and many other hotels are of this type.

29 fashion and art crossover design hotels are carefully selected and collected in this book. We hope that this book can provide a panorama of crossover design hotels for readers. For the sake of helping the readers to master a clear sequence of thought and frame of this book, we present the content of the book in six parts. They are as follows:

### Hotels in Fancy Dress

Hotels collected in Part One are all dressed up with fresh and colourful clothes by fashion designers. Revival, colour, art, dream and originality run throughout the design. As a result, hotels become the dreamland of these designers.

### Fashion Brands Hotels

Part Two introduces some famous fashion design hotels, including top-end fashion brands such as Armani, Versace, Casa Camper, Ferragamo, Tcherassi, etc. The charm of such hotels is that they make guests surrounded by famous brands they love. After all, the most direct way is to inject the existing brand style and elements into the process of the hotel interior design. Many designers advocate that fashion forward hotels can effectively convey the philosophy of brands at a broad level. Moreover, designers convert their demands into a

lifestyle to influence people.

### Timeless Classic of Fashion

All the hotels collected in Part Three present a typical design style admired by fashion designers. Usually, when designing fashionable clothes they absorb the cream of the crop of classic works and then make innovations. This style is true of the hotel design. Boasting a long history and traditional brand effect, Claridge's, Alma Schlosshotel inn Grunewald and The Beverly Hills Hoteland Bungalows are reborn via the involvement of fashion and in the meanwhile, fashion finally becomes timeless in the classic decorations of hotels.

### Hot Destinations of Celebrities

Part Four introduces the fashion hotels some stars and social celebrities often visit. In this part, several hot destinations are explored. With steadfast personal faith and pursuit, these fashion designers design fascinating settings for revelry and holiday.

### Hotels: Painting Rooms

Part Five displays the hotels possessing art elements. These hotels become the workshops of the artists where traditional patterns are broken and rules are redefined. Artists create the most inspiring Castalia through painting and sculpture in spaces.

### Hotels as Art Works

Part Six introduces some hotels featuring artworks to enhance interior design atmosphere. An art piece is a small part in hotel design. However, these hotels improve the level by using art pieces and artworks that become the protagonist for creating atmosphere.

Crossover design hotels are "the star" born as design hotel developed to a certain degree. It enhances the leading role of designer in hotel design. The personality and concept becomes the theme in crossover design hotels. At the same time, the premise of a successful crossover design hotel depends on the designers' personal charm and infection. The birth and upsurge of the crossover design which satisfies the designers' demands is by no means accidental. In fact, the crossover design is anything but a non-conventional approach. All the projects collected in this book are the mirror of a breakthrough made by designers in a broader and practical domain. The participation of fashion designers and artists lets ordinary guests experience the charm and in particular provides professionals with creative inspiration. When fashion fades away with the passage of time and art works pass into silence, the extraordinary feeling aroused by these crossover design hotels is endless, which is also the profound significance of the crossover design.

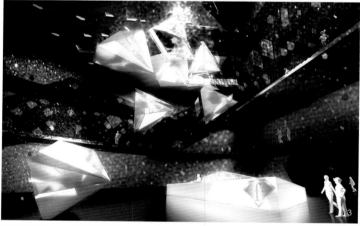

跨界这个词听起来总是让人琢磨不透，带着一层面纱。在各个领域出其不意的来一次现身，都吸引着看客去观摩，去研究。从最初，不同学科的融合影响了许多新学科的诞生，例如在文学界，女性主义的思想与叙事学的跨界，形成了一个新的领域女性主义叙事学；而在音乐研究领域，各种类型音乐的融合更是比比皆是，例如吸收了摇滚乐和爵士乐的元素而形成的节奏布鲁斯，如今更是电子乐与其他音乐类型融合的黄金时代。在这里，我们谈到的是另一种在设计领域当中的跨界。

设计界从不缺少跨界，因为在这里，通过艺术形态的拼接能够更直观的实现创作者的表达欲望，而设计领域的跨界产生的效果更是令人惊叹。让跨界在设计界如此的受欢迎的另一个原因，这与"创意"一词是分不开的。德国的平面设计大师马蒂亚斯将如今的设计含义直接与"跨界设计"等同。首先他认为创意就是将两个看似无关联的事物结合起来。如今复杂的设计市场需求，很多情况是通过不同知识领域的涉猎达到满意的效果的。这也正是跨界设计的价值所在，让不相干、矛盾甚至对立的元素擦出灵感的火花和奇妙创意。

跨界酒店的诞生让普通游客也能体会到跨界设计的魅力。它起源于设计酒店的设计浪潮，这类酒店最能透出设计的前沿趋势。从设计酒店的诞生之日起，就逐渐让人们意识到酒店是一个可以激发灵感，体验惊奇的地方。设计酒店强调以设计为主，通过定制化和个性化的室内设计，吸引着高品质的客人。这慢慢的形成了一种"体验经济"，人们想在下榻的酒店体会到更多，而不是提供一张床和一餐饭的临时住处。时尚爱好人士想在旅行时也让钟爱的品牌包围自己的生活，而艺术家更像要能够提供给他们更多灵感的住宿空间，还有那些追求品位和新奇的游客，这就促成了时尚和艺术酒店的诞生。这些跨界酒店可以说是设计酒店的延伸，成为了这些艺术形式的一种新载体。

设计瞬息万变，这与时尚界的快速更新和艺术品创作的不定性相通，因此室内设计与时尚、艺术的联姻顺理成章。当今的跨界酒店基本存在这两种形式，一种是时尚跨界，另一种是艺术跨界。最炙手可热的要数那些被名人追捧的时尚跨界酒店了。时尚和建筑业内的跨界其实已有先例。在最初，一些时尚设计师注意到设计服装与建筑结构的相通性。克里斯汀·迪奥曾经的得力干将皮埃尔·巴尔曼，也是高端时尚品牌巴尔曼的创始人曾说过，"衣服是一种流动的建筑"，他也成为在服装设计中受到建筑结构影响的第一人。此后，在巴黎世家（Balenciaga）和皮尔·卡丹（Pierre Cardin）的设计中都能看到外形硬朗，呈几何形的服装，时装界从此在结构、材料和廓形这些方面融入建筑元素，开启了建筑形式的探索。另一方面，一些建筑师也在作品中尝试以时装视角演绎建筑，例如赫佐格与德梅隆建筑事务所，在阐述他们的建筑理念时，曾明确表示自己受到服装设计用料的启发；Office dA事务所在韩国设计的办公楼采用服装制作中的打省、压褶的处理方法处理结构和材料。这些受到服装设计影响的建筑也为建筑的多样性和方向做出了贡献，表明建筑与时装界对话的可能。

进入到最近几年，时尚界和建筑界的交集集中在了时尚设计师设计酒店室内的热潮中。阿玛尼、范思哲、米索尼和宝格丽都开始各自建立同时尚品牌的酒店，这立即为跨界一词蒙上了一些商业的色彩，凭借前人跨界探索，他们有理由也有能力在跨界的商业之路上创建一座时尚丰碑。时尚业内，广泛涉猎一直是时尚品牌实现品牌影响力和效益的一种方式。通常一个稍有影响力的时尚品牌除了设计服装，还会涉及香水、箱包、鞋包乃至家具和家居用品的设计。另外，近年受到金融危机的影响，也迫使这些品牌拓展他们的设计领域，而酒店在具备一切东风的前提之下，自然成为了下一个目标。神奇的是，这些设计师是如何将时尚元素和室内设计相联系的呢？究其根本，在为人设计服装和为一间屋子设计室内装潢是有相通之处的。时尚界的多面手托德·奥尔德姆，他曾把设计室内看作给房间穿衣服，认为在给房间穿戴这方面，时装设计师能设计出更加赏心悦目的环境，并让置于其中的客人显得时尚。

另一类广受关注的跨界酒店是艺术跨界酒店。这类酒店目前主要集中为两类，一类酒店由酒店管理者组织、邀请艺术家设计酒店的室内装潢，多数情况，这类酒店的业主或者管理者就是艺术爱好者或是艺术家本身。例如位于多伦多的格拉斯通酒店，每一间客房都由一位艺术家原创设计，各家无法复制的艺术灵感交汇于此。同样，在新加坡的新大华酒店采用相同的模式复制。另一些酒店被设计师变成了艺术博物馆，在酒店的各处摆放着著名艺术家的作品，这让客人可以设身处地的生活在艺术品营造的氛围之中，还有机会更亲近地欣赏和品鉴这些艺术品。这类酒店更受到艺术家们的欢迎，目前在欧洲已经有专门为艺术家提供创作空间的艺术酒店。在中国，一些同类酒店也出现了，例如上海的斯沃琪和平饭店艺术中心（Swatch Art Peace Hotel）和北京格瑞斯酒店(Grace Beijing Hotel)。

在这本书中，我们精心挑选了最具代表的29家时尚和艺术跨界酒店，希望能通过这些为读者全面展示这类酒店的风貌。为了达到让读者能够有个清晰的脉络，我们将全书分为六个部分。

第一部分为"穿奇装异服的酒店"，这部分的酒店被时尚设计师的斑斓"服饰"打扮了起来，复古、色彩、艺术、梦境和创意贯穿在酒店的设计之中，酒店俨然成为了设计师们的梦想之地。

第二部分"时尚品牌的酒店业入侵"，主要介绍了已经成为了著名酒店品牌的时尚品牌酒店，这些品牌包括阿玛尼、范思哲、营地之家、菲拉格慕和切拉西。让旅行沉浸在自己喜爱的时尚品牌之中，这是这类酒店的吸引力所在。究其根本，最直接的方式就是将已有的固定的品牌风格和元素融入到酒店的室内设计中。很多设计者认为这类酒店能够更宽而有效的传达品牌哲学，他们想将自己的追求转变成一种生活方式去影响大众。

第三部分命名为"时尚演绎的永恒经典"，这部分的酒店体现了时尚设计师推崇的一种设计风格。时尚设计师在设计服装作品时，通常是在经典的作品中吸取灵感并加以创新，这在酒店的设计中仿佛也沿用了这一点。克拉里奇酒店、阿尔玛格吕内瓦尔德皇宫酒店和贝弗利山庄酒店及别墅，这三家具有历史和传统口碑的酒店经由时尚之手孕育出了第二次生命，而时尚也在经典的装潢中得以永生。

第四部分"名人们的时尚聚集地"，介绍了一些明星和社会名人经常光顾的时尚酒店。在这部分，我们细致探索几家世界各地的潮店，时尚设计师们秉持各自的个性信念和追求，设计出最吸引人眼球的狂欢、度假场所。

第五部分"酒店内的绘画空间"，呈现了几家带有相当艺术元素的酒店，这些酒店俨然成为了艺术家们的创作室，在这里模式被破解，规则也被重新定义。艺术家们在空间内用绘画和雕塑等形式创意出最能激发灵感的源泉。

第六部分"酒店艺术品"，主要介绍了一些通过艺术品来加强室内设计氛围的酒店。艺术品在一般的酒店只是装饰部分的一个小环节，而这里介绍的几家酒店将艺术品的运用提升到另一个层次，艺术品成为了制造氛围的主角。

跨界设计酒店是在设计酒店发展到一定程度时产生的"明星"。它加强了设计师的在酒店设计中的主导地位，设计师的个性特色和价值取向将成为酒店设计的主题，而这类酒店成功的前提正是这些设计师本身的个人魅力和感染力。跨界设计的热潮不是一种偶然的现象，观察各领域的跨界，它们的诞生满足的是设计者的设计诉求。其实跨界设计不是一种目的，而是创作的一种非常规手段。我们在这本书中了解到酒店设计中的跨界现象，是设计师们在一个更广阔的设计实用领域创作出的一次突破。时尚设计师和艺术家们的参与，让普通的客人能体验到设计的魅力，更能让业内人士获得创作的灵感。当时尚成为过去时，当艺术作品被人遗忘，这些魅力酒店带给人们的非凡体验却不会成为过眼云烟，这也正是跨界设计的深远意义所在。

1. FTV is the number one fashion media brand in the world. World-class architecture and interior design firm dwp was invited to transform the glamour, fun and creativity of the F brand into sensational architecture and interior spaces. This led to the birth of the F Hotel concept, which includes a 60-storey tower and a resort, inspired by the FTV brand. The image shows the concept for all day dining in FTV hotel.
2. The concept for lobby in FTV hotel
3. The night club concept in FTV hotel
4. Silver guestroom concept in FTV hotel

1. 时尚电视台（FTV）是世界顶尖级的时尚传媒品牌。世界级建筑和室内设计公司dwp将FTV品牌充满魅力、娱乐性和创造力的特质融入到建筑和室内设计中。受到FTV品牌的灵感启发，FTV酒店的设计理念随之诞生，一座60层的酒店塔楼即将诞生。图片为FTV酒店全日餐厅概念图。
2. FTV酒店大堂概念图
3. FTV酒店夜总会概念图
4. FTV酒店银色客房概念图

# Hotels in Fancy Dress
穿奇装异服的酒店

Hotels collected here are all dressed up with fresh and colourful clothes by fashion designers. Revival, colour, art, dream and originality run throughout the design. As a result, hotels become the dreamland of these designers. The three hotels designed by Christian Lacroix are the fruits of revival, colour and visual tramp, the magic of Lacroix. Comparatively, Frank Moschino is a fairy tale author. In his representative work Maison Moschino, sweet dream is created by surrealistic techniques. Martin Margiela, a master of deconstructivism, exercises his talent in the hotel design by leaving margin and makes the guests understand his aesthetic thoughts little by little. The enchanted Hotel ORiginal by cutting-edge fashion designer Stella Cadente is a new story inspired by some famous fairy tales and poetry.

酒店被时尚设计师的斑斓"服饰"打扮了起来，复古、色彩、艺术、梦境和创意贯穿在酒店的设计之中，酒店俨然成为了设计师们的梦想之地。克里斯汀·拉克鲁瓦设计的三家酒店是他用三个元素施展出的三个魔法，复古、色彩和视觉陷阱是他符咒之源；相比较而言，弗兰克·莫斯奇诺则是一个童话的作者，在他设计的莫斯奇诺酒店，梦境用超现实主义的手法创造出来；解构天才马丁·马吉拉让他的天赋在酒店设计领域发挥光彩，室内设计通过艺术留白的形式让客人慢慢回味、领悟他的美学思想；充满魔幻色彩的原创酒店，是新锐时尚设计师斯黛拉·卡丹特取众家之长拼接而成的新故事。

# Endless Revival of Old Things, Fashion Concept in Antique Building

让旧事物无休止的复兴,古董建筑中重新孕育的时尚思维

Christian Lacroix, who is a famous French fashion designer, is one of the famous design masters that are activated in Paris haute couture. Christian Lacroix has got the Gold Thimble award which is known as "Oscar" award in the fashion world, the prize of "The Most Influential Foreign Design" in the American fashion designer association awards, and also the reputation of "Paris Conqueror" in the fashion world.

"Endless Revival of Old Things" is the slogan of Lacroix's fashion design. His infatuation to antique can be found in his works. He began to study the clothing of 17th when he was a student. During his 20 years of fashion design, Lacroix developed a hobby of collecting antiques and the inspiration from these collections can be reflected in his works. Complicated embroideries, stacked laces, exaggerated jewelleries, deep colours, and the detail's handling of diamond feathers are as if to bring people back to the Gothic Age.

Just from the view of its location, Hotel Le Bellechasse meets Lacroix's love to the design restoring ancient ways. The street the hotel located in joints the two districts that symbolise the Paris aristocrat district and art district. The whole street can remind people of the Orsay Museum in the 19th century. And the antique furniture in the hotel verifies Lacroix's love to the old things. He displayed the furniture respectively in the suites of different themes such as "Revenge", "Saint Germain", "Tuileries", "Mousquetaires". These guestrooms which have historical marks were conceived again with bright colour, exaggerated figure and fashionable thought, and given a new life with the antique furniture.

克里斯汀·拉克鲁瓦是法国著名的服装设计师,是活跃在巴黎高级女装界的著名设计大师之一。拉克鲁瓦先生曾获得有时装界"奥斯卡"奖之称的金顶针大奖,以及美国时装设计师协会的"最具有影响力的外国设计"大奖,同时在业内还有"巴黎征服者"的美称。

"让旧事物无休止的复兴"是拉克鲁瓦时尚设计的标语。在他的作品中,能看到他对古董衣的迷恋。在学生时代,他就开始对17世纪的服装进行过研究。在从事时装设计的20年间,拉克鲁瓦养成了搜集古董衣的爱好,在他的作品中,也能看出从这些藏品中摄取灵感的影子,繁复的刺绣,层叠的蕾丝,夸张的配饰,浓重的色彩,钻石羽毛的细节处理,仿佛把人直接拉回哥特时代。

这家柏歇斯酒店,单看地理位置就满足了拉克鲁瓦对于复古设计的喜爱,酒店位于巴黎贵族区和艺术区的两大区域之间,整条街道让人联想起19世纪的奥赛博物馆,而走进酒店里面的古董家具正印证了拉克鲁瓦对旧事物的喜爱,他让这些家具分别"陈列"在不同的主题套房,"复仇者"、"圣日耳曼"、"杜乐丽宫"、"火枪手"这些带有历史印记的客房被用艳丽的色彩、夸张的图像和时尚的思维重新孕育,更赋予了这些古董家具以新生。

# Hotel Le Bellechasse

柏歇斯酒店

**Completion date:** 2007
**Location:** Paris, France
**Designer:** Christian Lacroix, Jean-Luc Bras, Emmanuelle Thisy, Anne Brugière Peyroux
**Size:** 34 rooms
**Photographer:** Hotel Le Bellechasse

完成时间：2007年
项目地点：法国，巴黎
设计师：克里斯汀·拉克鲁瓦，让-吕克·伯哈,艾曼纽·迪西, 安妮·布什耶·柏宏
规模：34间客房
摄影师：由柏歇斯酒店提供

Nearby the Orsay Museum, the "Bellechasse" converted into a delightful hotel, full of character, being altogether impressive and intimate, "dressed" by Christian Lacroix.

Ideally situated in the heart of the left bank, between the VIIth aristocratic and the VIth artistic districts, this elegant private hotel is now a real jewel case of "haute couture" where travelers and aesthetes will be able to meet.

Behind its classic appearance is hidden an outstanding personality. 34 rooms of character where designs, figures, colours and subjects are skillfully orchestrated and re-occur at the same times neo-classicism and Bohemian spirit.

Christian Lacroix explains: "In my opinion, a hotel must reflect the character of the locality it is standing in and should represent 'a travel within the travel' while giving its own interpretation of the town, the district, the street it is open to. This hotel provides accommodation to the tourists in search of culture and exhibitions as well as making them feeling 'at home' in a district essentially notorious between politics and business.

With regards to the street of Bellechasse, one must think in terms of the 19th Century aesthetics of the Museum of Orsay, with the old furniture of famous antique dealers, with the neo-classicism of architectures neighbourhood and also with the art galleries, contemporary furniture and an unconventional way of life which were, since always, the axis of Faubourg Saint-Germain whose charm, force and elegance come from this balance between tradition and audacity, past and present, upper middle class and dolce vita, symbolising for me the spirit of Paris."

On reflection the designer proportioned all these ingredients in function of all the places to be re invented. The reception mirrors this conception immediately by showing the blending between "the old and the new" and with just a hint of the future. The lobby-bar overlooking on to the patio-garden, plays on to the glass window and the mirror, the bright colours, the white mouldings, the wood storage and the blending styles of cosy furniture. The staircases are scattered with graphic images and covered with bright red carpet, the walk-ways are softened by round partitions, the ceilings are black and on the ground an exclusive carpet which the designer drew with strokes of black-ink on to a white back-ground. The colour allows giving rhythm distinguishing floor levels from each other. Generally monochrome, as well as camaïeu, sometimes coordinated to printed wall-paper.

Playing with a palette of ambiences and influences, perpetuating a skilful balance between audacity and tradition, Christian Lacroix created 7 universes of diversity throughout the rooms: PATCHWORK, AVENGERS, SAINT-GERMAIN, TUILERIES, MOUSQUETAIRES, JEU de PAUME, QUAI d'ORSAY

On the ground floor, the conservatory opening onto the patio with spaces of activities blending with the pastels and dark lacquer from the cosy lounges, while the rooms on the upper floor are more "futuristic" under the upside-down hull of a ship formed by the roofs. The hotel Le Bellechasse is the result of a skilful alchemy. The Baroque inspiration of the fashion designer is mixed with the influences of a highly historic district. The final result is a Parisian touch typically "left bank". A mixture of shapes and colours makes it most attractive.

坐落于奥赛博物馆附近的柏歇斯酒店经由法国时尚设计师克里斯汀·拉克鲁瓦之手重获新装，新的柏歇斯酒店整体设计明亮欢快，别具个性，令人印象深刻且不失亲切之感。

酒店地理位置优越，坐落在塞纳河左岸的中心区，衔接着第七贵族区和第六艺术区。这座优越的私人酒店如今是一座被高端时尚装点的珠宝盒，聚集了来到这里的旅行家和美学家们。

经典的建筑外表下隐藏着超凡的个性设计。在34间主题客房内，设计、图形、色彩与物品和谐的编排在一起，诉说着新古典主义与波西米亚风格的精髓。

设计师克里斯汀·拉克鲁瓦说："在我看来，一座酒店必须反映出它的特色并且会展现出'在旅行中旅行'的特色，与此同时能给出当地城镇、区域、街道的特色。在这个以政治与商业闻名的地区，这座酒店提供给旅行者发掘文化的机遇以及如家般的旅行感受。"

谈到柏歇斯酒店的街道，人们一定会想到有关19世纪奥赛博物馆一带象征着巴黎精神的美学表象，会想到博物馆内部著名古董商的旧家具藏品，会想到博物馆附近的新古典主义建筑，也会想到街上的艺术画廊，内部的现代派家具，以及圣日耳曼新区不变的非传统生活方式——在传统与革新，过去与现在，上中产阶级与叛逆派之间追逐的一种有魅力的，强烈且优雅的生活状态。

以下剖析这些美学元素在酒店各部分是如何应用的。夹带着一些未来元素，接待前台最直接的体现出将"新与旧"柔和的感念。大堂酒吧遥望露台花园，酒吧内的玻璃窗、镜子、明亮的色彩、白色的装饰性线条、木质储藏间以及混合风格舒适家居相映成趣。楼梯铺上了红色的地毯，且墙上随意的装饰了抽象画；走廊的天花板是黑色的，而地上铺设了一条

由克里斯汀·拉克鲁瓦亲手绘制的白底黑墨写意地毯。酒店每一层的走廊都用不同的色调与画作区分，表达出韵律感。这些画作以黑白为主，也有些是单色画，有时也搭配印好的墙纸。

用别具氛围与影响力的调色板创造，永远在传统与创新之间畅快的游走，克里斯汀·拉克鲁瓦利用这些酒店房间创作出7个不同的世界。这些房间被命名为：拼接画、复仇者、圣日耳曼、杜乐丽宫、火枪手、美术馆、奥赛码头。

酒店一层的温室直通露台，带着活动空间的露台与装点着水粉与深漆的舒适休息室连通，而顶层的客房设计更具未来主义，整体被置于一个貌似颠倒过来的船型屋顶之下。

柏歇斯酒店仿佛是一场高超炼金术炼成的精品。时尚设计师受到巴洛克风格的灵感启发，混合了当地的历史影响。最终见到的是一个典型的左岸巴黎风格酒店，一个混合了奇形怪状，各异色彩的最吸引人眼球的酒店。

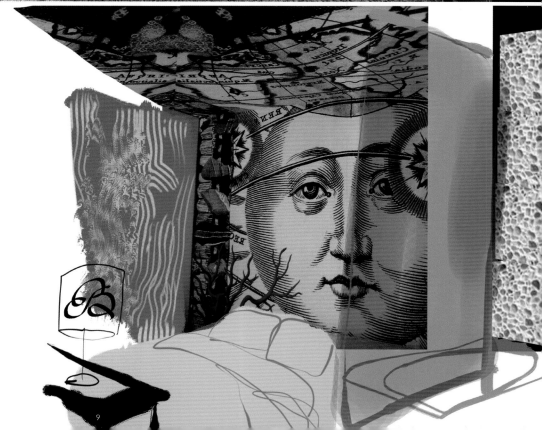

1. Wall in Discovery Room
2. The bedroom in Privilege Room
3. Sketches for Priviledge Room
4. The detail on the wall in Privilege Room
5–6. The detail on the wall in Original Room
7. Fantastic breakfast room
8. The overview of Original Room
9. Skctches for Qriginal Room
10. The desk in Original Room
11. The bathroom in Original Room
12. The artwork in breakfast room
13. The detail on the wall in Original Room

1. 发现客房的墙壁特写
2. 特权客房的卧室
3. 特权客房手绘图
4. 特权客房的墙面特写
5、6. 原创客房的墙面特写
7. 梦幻的早餐室
8. 原创客房全景
9. 原创客房手绘图
10. 原创客房内的书桌
11. 原创客房内的浴室
12. 早餐室内挂着的艺术品
13. 原创客房墙面细节

14. A corner in Discovery Room
15. Original sketches for Discovery Room
16. The bedroom and bathroom in Discovery Room
17. A corner in Privilege Room
18. Impressive room number brand
19. Original Room
20. The bedroom in Original Room
21. The Original Room with creative paintings on the wall and ceiling
22. A corner full of mystery in Discovery Room

14. 发现客房的一角
15. 发现客房手绘图
16. 发现客房内的卧室与浴室
17. 特权客房的一角
18. 醒目的房间号标志
19. 原创客房
20. 原创客房内的卧室
21. 原创客房的墙和天花板上绘制着充满创意的绘画
22. 发现客房内充满神秘色彩的一角

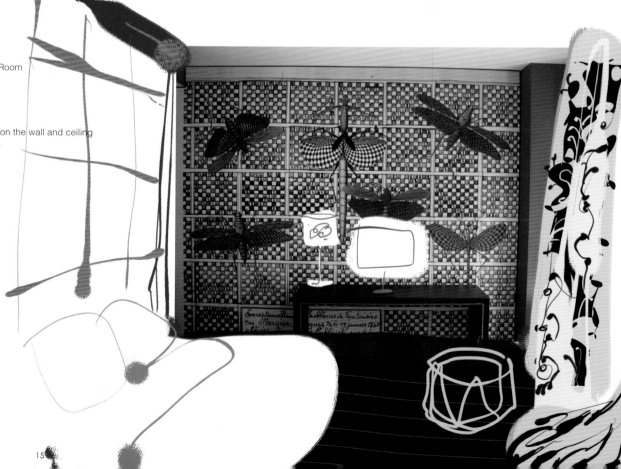

# A Colourful Time Created by the Palette Master

调色大师的彩色时光

In 1987, the clothing brand "Christian Lacroix" was founded after his name. During 20 years, he continuously presented the fashion "drama shows" to the fashion world, which has got hot pursuit in the world clothing, especially in 1980s when luxury is pursued. The most typical clothing style of Christian Lacroix is: gorgeous in modelling, beautiful in colours, and full of royal luxury and dramatic influence.

During 20 years, to infuse the South French and Spanish styles in his works is the consistent source of inspiration for Christian Lacroix. He was honoured "palette master". Proud pink, gorgeous orange, deep orchid, mysterious purple are contradictorily interweaved together, which becomes one of the features of Lacroix fashion. The love to colour comes from his native complex. Christian Lacroix was born in Arles, which is in South France and is adjacent to Spain, where these colours are common. And for example, gorgeous patterns on the clothing, blossom lace on the arm, sequins and bowknots adorned everywhere, all of them can show the Spanish royal luxury.

Hotel du Petit Mounlin is called by Lacroix "a colourful time" that is the mutual blending, attracting and collision of different colours under the collage and integration of different kinds of elements. For example, the reception of almond green combined with wine red, ornaments of bronzing black, frames of fluorescence white, figures of colourful flowers, paintings of deep purple, light of yellow and pink. The most attractive one is the wallpaper in the guestroom. From old golden to fluorescence green, it is like the source of the colour blooming, from which any colour can track.

克里斯汀·拉克鲁瓦在1987年创立了以自己名字命名的服装品牌"Christian Lacroix"，品牌存在的20年间，不断的为服装界敬献一场场时装"戏剧秀"。尤其在追求奢华的80年代受到世界时尚界的热力追捧。克里斯汀·拉克鲁瓦的最典型的服装风格为：造型华丽，色彩艳丽，充满宫廷式的奢华和戏剧般的感染力。

将法国南部风情和西班牙风格融入到作品中，是克里斯汀拉克鲁瓦20年来始终如一的灵感源泉。克里斯汀·拉克鲁瓦又被誉为"调色大师"，傲人的桃红，绚丽的橙黄，深邃的幽蓝，神秘的炫紫对比辉映地交织在一起，成为拉克鲁瓦时装的一大特色。对于颜色的喜爱，这来自于拉克鲁瓦的故乡情结，设计师出生在法国南部的阿尔斯，邻近西班牙，这样的色彩在这些地区是会经常见到的。再比如，时装上华丽的图纹，衣袖上绽放的花边，随处点缀的珠片和蝴蝶结，西班牙式的宫廷奢华也可见一斑。

这家小磨坊酒店被拉克鲁瓦称为"一段彩色的时光"，在各种元素的拼贴组合下，是各种颜色的互相融合，吸引与碰撞。杏仁绿配酒红的接待区，黑色泛金的装饰品，荧光白色的框架，彩色的花朵图纹，深紫色的漆绘，黄色和粉色的灯光，而最吸引人的是客房内的壁纸，从古金色到荧光绿，仿佛是颜色盛开的源头，任何颜色都有迹可寻。

# Hotel du Petit Moulin

小磨坊酒店

**Completion date:** 2011
**Location:** Paris, France
**Designer:** Christian Lacroix Design / Bastie Architect
**Photographer:** Christophe Bielsa

完成时间：2011年
项目地点：法国，巴黎
设计师：克里斯汀·拉克鲁瓦设计公司，巴斯蒂耶建筑事务所
摄影师：克里斯托弗·贝尔萨

"When Madame Murano, Mr. Nourry asked me to collaborate with them on their hotel projects last fall, I had the impression of rediscovering an erstwhile childhood dream left by the wayside. A dream of living in a hotel, building a decor, day after day, in the 'colours of the times', putting ambiences together into volume and space and not only on paper or through fashion collections. I was really convinced by this rather personal and intimate hotel concept, so different to the prevailing lookalike style of the big classic hotel chains."

This first project (three are to follow) is located on the corner of the Poitou and Saintonge streets where two old buildings have been joined together and refurbished by the Bastie architectural partnership.

Not so long ago, the first building housed a bakery. And not just any bakery. It was the oldest one in Paris, dating from Henry the 4th. The local legend is that Victor Hugo used to come and buy his bread there. The 1900 frontage is listed as part of French heritage by the Historic Monuments and the shop signboard has been conserved, together with the one on the neighboring hotel that can be seen on some old post cards, as has the original woodwork's black colouring slightly highlighted with gold, visible under the exterior glass fixtures. The very suave interior decoration of the old bakery, going right up to the ceiling has, of course, been restored and creates a "Venetian" style setting for the reception area in almond green with wine-coloured ruffled taffeta drapes. The designer immediately liked the slightly crooked and off-kilter perspectives, like the ones on the engravings of the time: the maze-like design of each of the floors, the fetal style refuge of the rooms and the new very functional spaces created while respecting the picturesque "old Paris" look of the listed elements, like the lovely 17th century wooden staircase left natural and enhanced by simple white pebble-dashed walls. All of these very special volumes dictated one single idea that seemed to designer to be coherent with such a place and location: each room really had to be very different, personalised and individually decorated in relation to its orientation, the height of its ceiling and its location within the heart of the hotel. And even the view. But most importantly, each room had to relate the beginning of a story to be finished by the guests themselves. Without, of course, falling into the artificial look of a pretend home or an ersatz apartment. Nor into the uniform concept of the hotel room.

Everyone was prepared to trust the designer and to follow me in this rather non-academic direction. Anne Peyroux has helped me to concretize this utopia.

Having left Saint-Germain to live in this area, the designer knows just what people are looking for here, how to live here, how to feel at home here, even when passing through. The Marais district has a very special and distinct personality of its own. It is rich in various very paradoxical facets, whether you are interested in contemporary art, fashion, vintage or not, or quite simply a certain life style somewhere between ancient stones and modernity - a mix of every trend. It is all these contrasts that the designer has endeavored to translate into seventeen ambiences corresponding to each of the seventeen rooms, like seventeen ways of experiencing the Haut-Marais. Beams or cement, antiquated wallpaper or gaudy fabrics, sober canvas fabrics or leather, contrast with wooden floors and tiles, carpeting and ceramics. An entire collection of contemporary lighting units,

60 seating units covered in brocade or velvet, or a little bit of fur and some period furniture upholstered in very bright colours and graphics, constitute the vital thread linking this medley of different worlds and finish by setting the tone for this patchwork of moods with panels of giant collages on the walls or windows. Consequently, the designer pass from a "rustic" Marais in toile de Jouy to a more "Zen" or "design" Marais or from an historical Marais in damasks to a more playful Marais. Via lawn green corridors with black lacquered doors framed in optical white on a polka dot carpet background, we proceed from decors designed with pilasters, cornices and consoles "as a joke", modern bathrooms in slate, faïence or cement warmed with Venetian mirrors, ceramic kaleidoscopes or panoramic wallpapers, from baroque, rococo or "Couture" rooms, from 21st century spareness to a wink at nostalgia, from masculine to feminine, from north to south, from flowers to stripes, from antique gold to fluo green. Likewise, the designer pass from the slightly "cake shop" reception to a more somber and rich egg-plant lacquered salon and from the elevator decorated with collages of antique engravings to the bar, treated like a local street café that combines the zinc counter and "Edwardian" furniture with "scrap-book" walls and 60s seating in shades of yellow and pink.

It's like couture, where the harmony is created from a puzzle of inspirations, where the feeling of the moment is nourished with elements from the past, where modernity lives in the tradition of the present.

1. The magnificent background of the bed in Superieure Room
2. The colourful reception
3. The delicate mirror in the bedroom
4. The artistic corner in Superieure Room
5. The bedroom in Junior Suite
6. The overview of Deluxe Room
7-9. The details of guestrooms in the view from interested angles
10. The overview of Superiure Room
11. Confort Room
12-13. The details of Deluxe Room
14. The Deluxe Room full of lavish colours and patterns
15. The details of the bedroom in the mirror
16. The view from stairs
17. The luxurious bathroom

1. 高级客房内的绚丽床头背景
2. 色彩汇聚的接待区
3. 卧室内精致的镜子
4. 高级客房内充满艺术气息的一角
5. 普通套房的卧室
6. 奢华客房全景
7-9. 从一些有趣的角度看客房的细部
10. 高级客房全景
11. 舒适客房
12、13. 高级客房细节
14. 装点着丰富色彩和图案的高级客房
15. 镜子中的卧室细节
16. 从楼梯看客房
17. 奢华的客房浴室

"当穆拉诺夫人,诺里先生邀请我设计他们的酒店时,我的脑海里闪现出我儿时的一个梦境。在梦里,这是一段彩色的时光,我住在一家酒店,日复一日装扮着它,将不同的布景放入这个空间,不仅仅是铺上壁纸或者放入时髦的收藏品。这样个性又亲切的酒店设计概念让我深深的难忘,它与现在普遍的大型经典酒店看起来完全的不同。"克里斯汀·拉克鲁瓦谈到小磨坊酒店设计时说。

小磨坊酒店坐落于普瓦图街与圣东日街之间的交汇处,这里原是两座建筑,现在经巴斯蒂耶建筑事务所改造为一座整体建筑。

在改造之前,这里曾经是一家面包店,而且不是一家普通的面包店,这里曾是巴黎最老的面包店,可以追溯到亨利四世。传说,维克多·雨果曾经在这里买过面包。店面已经被列为法国历史遗迹,店铺的告示板也被保留了下来,与邻近的酒店一样,它黑色略微泛金色的木制装饰品,透过窗户清晰可见,这样的场景还被经常印在明信片上。古旧的面包房内,重新装修过后,文雅的室内装饰遍布整个空间。杏仁绿色的接待区用酒红色的带褶皱塔夫绸窗帘装饰,布局营造出威尼斯风格。设计师尤其喜欢酒店内稍有弯曲又不平衡的视角,像是时光的雕刻艺术品:迷宫般的楼层,避难所一样的房间,展现巴黎旧貌的如画般的新型实用空间,以及17世纪风格的木质楼梯在简单白卵泥灰墙的映衬下显得非常自然。所以这些特别的空间展述了一个与酒店当地风貌十分一致的理念:每一间房间都是如此的不同,有个性,各自单独装饰,依照不同的朝向,房间高度,方位,甚至是视角,特别装饰定制。然而,最重要的是,每个房间都能开启一段故事,结局怎样就要靠客人们自己了。在这里,客人不会感觉来到了仿造的家或者仿制公寓楼中,也不会感觉酒店房间像是穿上了统一的制服。

在设计师克里斯汀·拉克鲁瓦设计师非学术性的指引,以及建筑师安妮·柏宏的辅助下,客人们来到酒店仿佛来到了一个新建的乌托邦。

对于那些刚离开圣日耳曼区而来到这里休息的游客,设计师敏感的捕捉到他们对这个地方的期望,他们想在这里如何起居,怎样才能让他们感到自在,甚至当他们只是路过这里他们又需要怎样的酒店体验⋯⋯这些都被设计师在酒店设计之初铺垫入酒店的设计之中。马莱区是一个非常特别且具有鲜明特色的地区。在这里蕴含着许多对立的文化,无论当代艺术、时尚、复古文化,又或者是混搭潮流——介于古老与现代之间的生活方式在这里都可以见到端倪。设计师努力将这些冲突的文化元素转化成17种主题,赋予酒店的17间客房以新的生命,客人们住在其中,用17种方式体验马莱区的上流生活。梁柱或水泥;古旧的壁纸或华丽的织物;沉稳的帆布或皮革面料与木质地板;瓷砖、地毯及瓷器都形成鲜明的对比。酒店内整体采用的现代灯饰,60个装饰有天鹅绒与锦缎的落地灯,铺着毛皮或亮色软垫的旧式家具以及图画,将这个混杂的时间连接到一起;墙上和床上的拼接艺术品、拼接画最后完成了酒店整体的拼接基调。

最后,设计师将乡土气息的马莱区改造成了设计感十足的马莱区,将一个穿着锦缎的历史马莱区改造成一个更富玩味的马莱区。穿过以绿色草坪为背景,沿着以下顺序会看到:黑色漆木、荧光白色框架、铺着波尔卡圆点地毯的走廊,接下来的是沿袭着布满装饰的柱子、飞檐和接待台,铺着板岩、装点彩釉陶器的现代派浴室,用威尼斯镜子、陶瓷万花筒和全景壁纸装饰的客房。客房内的壁纸最为鲜明,图案从巴洛克,洛可可到高端时尚,从21世纪的节俭风到怀旧主义,从男性主义到女性主义,从北方到南方,从花朵到条纹,从古金色到荧光绿。同样的,客人们可以从蛋糕店式的接待处到达更为昏暗、深紫色的漆绘装饰沙龙;可以从装饰着古董雕刻品的拼贴画的电梯到达街头咖啡馆式的酒吧,这里有镀锌的吧台,爱德华时代的家具,墙用书的碎片装饰着,60个座位在黄色和粉色的光影下忽隐忽现。

这一切都像是一场时装秀,在这里,和谐在灵感的谜团中产生,在这里,此时此感被过去的元素滋养,在这里,现代性在当今的传统中成长。

漫步游走在17个客房和私人酒吧之间,除了可以看到由拉克鲁瓦时装手绘的图样或挑选的画作之外,更多的是设想着马莱区的十七种各自不同的风情。当一切都是在法式优雅中,又肆无忌悼的同时,却惊见壁面上,北欧风格的壁纸和圆点点的毯子与装饰细节,正定定地在空间中没蔓延着⋯⋯这或许就是拉克华才有的胆识,每一段马莱区的风貌,在他手里,都是画笔上可供一层层涂上的油彩,"融合"的意图也许未必,"拼贴"或许才是他高招的真意。每一个楼层间,迷宫似的过道,搭配重新规划出来的功能区域,一整个概念无非就是要在"旧巴黎"的样式中,探寻出经典足以历时至今的部分。

# The Trompe l'Oeil by Art and Colour, The Birth of Neo-Braoque

艺术与色彩的视觉陷阱，新巴洛克的诞生

This third Hotel should be in the prolongation of the previous ones, without paraphrasing them; it keeps a family resemblance without showing it. The exceptional location of the Hotel Le Notre Dame far from being insignificant – at the intersection of all ages and all cultures – could inspire the designer another form of patchwork.

Christian Lacroix studied art history when he was a student. He loves painting and has profound cultivation in art. Besides luxury, a rich art breath can be experienced in Christian Lacroix's fashionable dresses which have strong feelings of painting, illusion and reality for absorbing the cultural essence of South France and Spain. The model who wears the elaborately hand-made dress is like the one who walks out of the painting. His design of hotel can also reflect his love to art's study and practice. In Hotel Le Notre Dame, guests can see the trompe l'eoil everywhere. It is a kind of art form that can only be seen in Paris and was embedded into interior design by Lacroix in different ways. He innovatively applied this kind of painting form to floor decoration such as the carpets and floor boards. What is more wonderful is that he can apply it to different materials. Moreover, combined with the mixed colours that Lacroix is skilled in, some famous paintings such as Portrait of Arnolfini and His Wife by Jan Van Eyck are selected to decorate a few guestrooms as a part of the collages background, which make the whole hotel feel mysterious and have a sense of the times.

That's why he liked to fancy a vaguely medieval style and find a contemporary expression far from the clinical coldness, which proposes a new Baroque, and lets the visitors travelling in their own imagination during their stay in Paris. Because Lacroix loves the feeling of being elsewhere in an almost dreamed place both in agreement and contrast with the city and daily life, closer to us because the vocation of travelling is to know oneself more deeply.

圣母院酒店是继柏歇斯酒店和小磨坊酒店之后，克里斯汀·拉克鲁瓦设计的第三家酒店。这家酒店可以说是前两家的设计延伸，是拉克鲁瓦设计的再次实践和诠释，圣母院酒店这个项目让拉克鲁瓦的设计具有了一种完整性。它独一无二的选址，介于巴黎彰显各个时代文化的街区之间，让设计师有了用拼贴画装饰酒店的灵感。

学生时代克里斯汀·拉克鲁瓦研修过艺术史，热爱绘画，有着深厚的艺术修养。在克里斯汀·拉克鲁瓦的时装中，除去奢华，也会让人感觉到浓郁的艺术气息，在吸取了法国南部和西班牙文化的精髓之后，有着强烈的绘画感，迷幻而真实，模特穿着用手工精心制成的时装仿佛是从油画中走出的画中人。对于酒店的设计也体现出他对于艺术研究与实践的热爱，在圣母院酒店中，客人会对那些无处不在的"trompe l'oeil"（视觉陷阱），这种只有在巴黎才能看到的艺术被拉克鲁瓦用不同的方式植入酒店的室内装饰当中，他创新的将这种绘画形式用在地面的装饰上，在地毯上，地板上，更奇妙的是用于不同的材质之上。此外，其中几间客房选用了名画作为拼贴画背景的一部分，例如扬·凡·埃克的《阿尔诺芬尼》，加之拉克鲁瓦擅用的颜色混搭，整个酒店显得既有神秘感有具有时代感。

这就是为什么设计师想象用模糊的中世纪风格进行一场现代式的表达，让酒店远离诊所似的冷漠感，也形成了一种新的巴洛克风格，让游客在巴黎停留的这段时间，用他们自己的想象力漫游。设计师所期望的是让游客能在一个梦想之地同时感受到与城市的日常生活相同和不同的两种感受，而来到圣母院酒店是了解巴黎的一种更有效的途径。

# The Hotel Le Notre Dame

圣母院酒店

**Completion date:** 2009
**Location:** Paris, France
**Designer:** Christian Lacroix
**Size:** 6 guestrooms
**Photographer:** The Hotel Le Notre Dame

完成时间：2009年
项目地点：法国，巴黎
设计师：克里斯汀·拉克鲁瓦
规模：6间客房
摄影师：由圣母院酒店提供

The Hotel "Le Notre Dame" is located between the cathedral Notre-Dame de Paris which gives it its name (and what a name!) and the studio where the painter Albert Marquet fixed forever, on the canvas, the grey-green-blue shades so typical of Paris and so subtle that the Seine offers all day long and throughout seasons. The Hotel Le Notre Dame has been a nice challenge to face for Christian Lacroix.

Beyond religions and soulless tourism, we can only be sensitive to such a privileged location in the heart of Paris and History. The Hotel Le Notre Dame could not ignore this aesthetic and spiritual aspect. That's why the designer did not want to make this place to be a monastery or a convent but a resting place which reflects – in the way of today – the wealthy centuries the district went through, always haunted by artists, scholars and students, an intimate place too.

The entrance to the Hotel on the quay is both sober and elegant with its black colour frontage, scalloped awnings, and the blue colour recesses; it draws attention without hype. But once he has crossed the threshold the traveller is immersed into the special atmosphere of this new Hotel through the compositions-collages on canvas framed with stones.

Instead of the traditional furniture, counter-furniture, chiné, technical, wood or metal, in contrast with a wall decor inspired by curio cabinets or shops of alchemists. The traveller is welcomed by a carpet printed with a huge motif of flowers and he can follow the baroque volutes in the lobby wide open on one of the most majestic view in the world: the quays of Paris, their booksellers, the Seine and the "vessel" Notre-Dame.

This angle space on the first floor, both lounge and breakfast room, is the perfect place to enjoy the landscape of the "Île de la Cité", to linger, to meet. On the walls a series of frescoes on canvas evoking "Lutèce" with an antique map of Paris and, at the back, still lifes "trompe l'oeil", black and white, and colour, nod to the opulence of past centuries. This space with seats dressed in coloured velvet, striped or embossed, has also a bar-counter entirely covered with mirrors reflecting the sky and light of Paris.

The visitor who will choose to walk up six floors will discover a staircase with colourful and graphic steps, walls hung with united velvet whose contrasting harmonies are extended into the corridors. The designer choose to cover the doors of a burning "trompe l'oeil" simulating an antique panelling on which, as a stencil, stand out the numbers of the rooms. The floor of the corridors is covered with a carpet backing of ethnic inspired embroidery Eastern. Thus, even the common areas are surrounding the clients with a cosy and warm, opulent and shimmering atmospheres, subtly playful.

For the rooms the designer have imagined six styles depending on their location and brightness: four for the rooms overlooking the street, one for the rooms onto the courtyard and the last one for the rooms under the roof. This ranges from classic to fantasy, with a fresco different for each of theses atmospheres. The bathrooms are all clad in stone or marble and, externally, with a false-wood box in various wood species. All this in order to give a personal note to the rooms.

圣母院酒店坐落在巴黎圣母院和画家阿尔伯特·玛盖的工作室之间，酒店也因此得名。酒店内的画布上到处是灰、绿、蓝，这在塞纳河是每天一年四季都会见到的盛景。对于拉克鲁瓦来说，设计这座圣母院酒店确实是一项挑战。

超越宗教，不只顾那些盲目的旅行者的看法，设计师不仅仅因酒店位于巴黎中心地带的优越地理位置和它的历史性说事。圣母院酒店的设计突出的是对设计师的美学观点和精神层面的表达。设计的初衷并不想把酒店装潢成现代的修道院，而是想把它设计成能够反映当地几个世纪以来的富饶景象，能够成为艺术家、学者、学生都喜爱光顾的酒店之一。

酒店的入口显得深沉和优雅，黑色的门脸，扇形的遮阳篷，蓝色的通道，这一切自然地吸引着游客们的目光。客人穿越石头墙的外立面，进入酒店就会沉浸在特别的氛围之中。

代替传统家具、前台、瓷器、木材及金属材料的是受到古董柜子和炼金商店装饰风格影响的装饰墙。迎接客人的是一条印着大花朵团的地毯，接着是巴洛克式的螺旋楼梯，开启的是世界上最壮美的风景，正是那些畅销书中所描述过的地方：塞纳河和圣母之舰。

客人走上六楼会发现一段色彩缤纷并装饰几何图案的楼梯，墙上贴着拼贴的法兰绒，楼梯与墙的这种反差对比形成了一种和谐，一直延伸至走廊。在酒店一层，酒廊和早餐室是欣赏这座城市景观的最佳地点，适宜放松休闲和聚会。墙上一系列的壁画搭配了一张巴黎地毯，让餐馆显现出一丝"卢特西"餐厅的味道。在空间的后方，生动的拼贴画将白色和黑色相互碰撞，联合其他色彩向过去的繁荣表达着敬意。空间内的座椅都铺着艳丽的法兰绒，上面不是印着条纹就是浮雕图案，同样，吧台上安装了几面镜子反射出天空和整个巴黎的城市灯光。

设计师受到古典镶板的影响，用绚烂的拼贴画遮盖住门板。走廊地面上铺着绣着东正教特色的图案的地毯底布。以此即使是普通的区域，围绕在顾客周围的也是惬意温暖，色彩丰富，充满熠熠星光，玩味凸出的氛围。

根据不同的方位和采光度，设计师设计出六种风格的客房：其中四间能够看到街景，一间在庭院之上，最后一间在顶层。他们从古典到梦幻，根据不同的氛围配上不同的壁画。所有的浴室用石头或者大理石打造，外部用不同的木材打造出木盒的造型。所有这些都令客房显得与众不同。

### 天使客房

这是第一间看得到街景的客房，混合了天鹅绒蓝色和花蕊的黄色，搭配了大马士革花梨木色窗帘和原木纹理的"视觉陷阱"艺术地毯，这些让房间看起来打破了传统。床的上方和周围贴着"后中世纪"的画作，仿佛天使正在客人的周围保护着，画作背景使用乡村手工织布和连着穗子的锦缎。整体房间有着"高级时代"的风格混杂着佛罗伦萨的特色。

### 夫妻客房

第二间临街的客房，地摊上的原木纹理的地毯被绘上"视觉陷阱"的地板所替代。酒瓶绿色搭配桃花心目天鹅绒与光滑的棕色窗帘形成反差。壁画围在床的周围，多了些"新文艺复兴"的韵味，这让客人感到受到凡·埃克的画作《阿尔诺芬尼》夫妻的保护，画作中的人在充满花朵的花园中，花朵用华丽的天鹅绒刺绣，让这些教科书上的图画焕然一新。

### 神圣花园客房

第三间客房里混合了"宫廷蓝色"和"金沙色"，让客房多了些田园风韵，地毯上绣着如石板路似的"视觉陷阱"拼贴画。花丛这回在这间英式花园的客房充当布景的角色，让房间透露出神圣的东方色彩。

### 祭司客房

铺着硬木地板，栗色和海军蓝色的天鹅绒，米黄色的锦缎，多了些古典感，这与画布上生机勃勃的壁画形成对比。画布上守望夜晚的祭司，装饰用的挂毯和锦缎让这类临街的客房自成风格。

### 1* The Angels

The first atmosphere "street" combines velvet navy blue and yellow "pistil" with the Damascus "rosewood" of the curtains while the carpet is printed with wood logs "trompe l'oeil" to break the traditional character of these smooth seate. Above and around the bed there is a composition "post-medieval" where primitive angels seem to protect the travellers' sleep on a background of rustic fabric and brocade embroidered with precious bunches. With something of Florence in the same mind "Haute Epoque" which the designer preferred for this project.

### 2* Spouses

For the second atmosphere "street", the wood logs on the carpet have been replaced by a gross floor "trompe l'oeil". The bottle green and mahogany velvet contrast with the glossy brown curtains. The fresco framing the bed, more "neo renaissance", puts the

traveller under the protection of the Almofini spouses of Van Yeck in a garden of bunches of flowers "petit point" embroidered lined with richly carved velvet that just up-date a few images from collage books.

### 3* Divine Garden

The third atmosphere where velvets are "royal blue" and "golden rust" is more rural despite the paving stone "trompe l'oeil" on the carpet. Because it seems that huge bunches of flowers are rising up from the bed on a background of English garden where some deities and oriental characters are hidden.

### 4* The Magi

Hardwood blocks on the floor, maroon and navy blue velvet, beige damask, more classical, contrasting with the most exuberant fresco on canvas where it is squarely shown that the Magi watch over the night, framed with tapestry and damask, characters and small collages coming from elsewhere for this last atmosphere "street".

### 5* The Courtyard

In contrast to these four rather opulent atmospheres the designer wanted one, more rustic and neutral, for these rooms which won't be cut off from the view because the designer anticipated that screens should rebroadcast what happens on the quay.

On the floor, printed pavement, a natural harmony for curtains beige and ecru broadband. Skin toro spotted on the walls, the seats or the bed with striped motives, kilims and arabesques brown, natural and chocolate. Above the bed sits a fireplace mantle and in a medallion on background of leather and lace, with giant carnations and a few scrapbook pages, a young girl, with downcast eyes, is watching.

### 6* Under the Roof

For the last two floors, very intertwined, the designer thought of a giant patchwork playing with all these bevelled surfaces. A hotter range combining rustic herringbone with scratches, blood and gold to the sepia coloured "toile de Jouy", skin mottled to golden wood. As for the canvas cover it is declined in several motives, black and white, juxtaposing giant lace and prints, architectural elements and furniture.

In each of these rooms a desk-like dressing table made of various wood species and shades according to the atmospheres, serves as the common thread running. The seats as well as bedside lamps or desk lamps change from one a scene to the other.

### 庭院客房

与之上四种色彩丰富的客房不同,庭院客房被设计的更为纯朴和中庸,因为不想令室内的装饰影响客人欣赏外面的海湾景色。

客房的地板上印着人行道的图案,宽大的窗帘是米黄色和淡褐色,显得自然而和谐。墙上用牛皮装饰,座椅和床印着条状花纹,搭配着基里姆地毯,到处是阿拉伯式的棕色和天然的巧克力色。在床上方是壁炉的地幔,背景是以皮革和蕾丝为材料做成的圆形浮雕,周围搭配康乃馨的图案和手绘本插页拼接而成的壁纸,画中一位少女用低垂的双眼注视着一切。

### 顶层套房

在最顶端的两个楼层,空间错综复杂,对于有坡度的地面,设计师用拼贴画装饰起来。这些画的图案包括乡村风格带划痕的人字形图案,红色、金色和深褐色的约依印花布图案,和金色木纹的斑驳图案。客房内的墙上贴各式各样图案的画布,色调为黑色和白色,上面并排印着蕾丝和图案、建筑元素和家具图像等。

在每一间客房都有用各种木材制成的书桌似的梳妆台,依据各自房间不同的氛围被制成不同的样式,可用于基本的梳洗打扮。客房内的座椅、床头灯和台灯也根据不同的客房有不同的样式。

1. The view from the balcony to appreciate Le Notre Dame
2. The detail of fantastic wall
3. The bedroom in double room
4. The detail of wall in double room
5. The double room with the Almofini
6. The angel theme room decorated with a composition "post-medieval"
7. The double room with the theme of magi
8. The lounge with colourful velvet chairs and frescoes
9. The bar counter covered with mirrors
10. The double room under the roof
11. The sofa corner in double room under the roof
12. The room under the roof with black and white graphic patterns
13. The twin room with the theme of courtyard
14. The door covered "trompe l'oeil" in the corridors
15. The colourful stairs
16. The unique basin
17. The corner with desk and bathroom in the guestroom

1. 在酒店的阳台上欣赏巴黎圣母院
2. 梦幻背景墙细节
3. 双人床客房的卧室
4. 双人床客房的装饰墙细节
5. 装饰着名画《阿尔诺芬尼》的双人床客房
6. 装饰着后中世纪风格名画的天使主题客房
7. 祭司主题的双人床客房
8. 布置着彩色法兰绒座椅和壁画的酒廊
9. 镜子组成的酒吧吧台
10. 顶层的双人床客房
11. 顶层双人客房的沙发一角
12. 装饰着黑白几何图案的顶层客房
13. 庭院主题的双人间客房
14. 走廊内的拼贴画装饰门
15. 色彩艳丽的楼梯
16. 独特的洗手盆
17. 客房内的书桌和浴室一角

# Surrealistic Sweet Dream

超现实美梦主义

With 20 years of history, Moschino is a famous fashion brand produced in Milan which is the fashion city in Italy. The design of its founder Frank Moschino is full of black humour that is ahead of fashion and surrealistic fantastic colour. The most impressive satire method had been used on Chanel. Moschino changed the classic suit of Chanel to a beggar's clothes by cutting the fringe and matching super buttons. It reflects his unrestrained and vigorous imagination and wizard design ability that brim with traditional view of fashion.

Moschino's pursuit for the game feeling is also reflected in his design of the Moschino Hotel. For example, the sheep sculptures in the guestrooms, the fly type pendants on the ceiling, and the king-size bed covered with huge red evening dress. Moreover, the lantern dress which is the representative work of Maschino brand is also applied in the decoration of the hotel. The striking decoration of the lantern dress in the lobby completely expresses the brand's surrealistic style and infinite imagination.

Another feature of the Moschino brand is its pursuit for the good and peace. In its fashion design, the smiling face of fresh yellow and the heart of brilliant red are the most commonly used brand mark decoration. It's the same to Moschino hotel. Each guestroom is endowed with a good and optimistic theme, such as "Alice in Wonderland", "Little Riding Hood". These themes of fairy tales can reflect that the brand advocates optimism.

莫斯奇诺（Moschino）是产于意大利时尚之都米兰的著名时尚品牌，有20年的历史。创始人弗兰克·莫斯奇诺以设计具有时尚超前的黑色幽默与超现实主义的幻想色彩著称。最让人深刻的一次讽刺手法用在了香奈儿身上，莫斯奇诺将香奈儿的经典套装剪破边缘，改装成乞丐装，另配上特大纽扣，他极具天赋的传统时尚视角能同时表现出品牌天马行空的想象力和鬼才设计能力。

莫斯奇诺对游戏感的追崇如今也体现在莫斯奇诺酒店的设计上，客房内的羊雕塑，天花板上的苍蝇型吊饰，以及用巨大的红色晚礼服覆盖的大床。另外，莫斯奇诺的品牌代表作灯笼裙也被运用在酒店的装饰中，在酒店的大堂，醒目的灯笼裙装饰彻底展示出品牌的超现实主义风格和无尽的想象力。

莫斯奇诺品牌的另一个特色是对美好以及和平的向往，在它的时装设计中，鲜黄的笑脸和艳红的爱心是最常出现的品牌标志装饰。同样的在莫斯奇诺酒店中，每一间客房都被赋予美好、乐观的主题，如"爱丽丝梦游仙境"、"小红帽"这些童话主题影射出品牌对乐观精神的倡导。

# Maison Moschino

莫斯奇诺酒店

**Completion date:** 2010
**Location:** Milan, Italy
**Designer:** Moschino
**Area:** 3000 m²
**Photographers:** Massimo Listri, Martina Barberini, Ake E:son Lindman, Henri Del Olmo

完成时间：2010年
项目地点：意大利，米兰
设计师：莫斯奇诺
规模：3000平方米
摄影师：马西姆·立斯特里，玛蒂娜·巴贝里尼，阿克·伊森·林德曼，亨利·德尔·欧尔摩

12 in Milan is the address of the new Maison Moschino hotel designed by the world-famous fashion label. By creating this hotel and adding the moniker "Maison", Moschino clearly wants to transform the repose of its guests into a fabulous dream in an intimate, personalised setting. "When I enter, I feel as if I'm at home. Like all familiar places, its warm, welcoming atmosphere makes you feel protected. I'd love to preserve it just as it is, as if it were an installation, but I know that, by nature, it is destined to change. And yet I believe it will never betray its origins. It is completely unlike other hotels, and it also has the courage to reveal the high-quality workmanship that built it. It's a special place, which is why it's called Maison Moschino."

With these words, Rossella Jardini, the Creative Director of Moschino, presented the fashion label's latest pet project: Maison Moschino. This "home and place of enchanted fairytales" opens the doors to its 65 rooms, its "Clandestino Milano" restaurant headed by acclaimed Chef Moreno Cedroni, its bar, its "Culti" spa, and its boutique for its official debut in Milan today.

Situated in a neoclassical building that was once the city's first railway station and is now at the heart of what is becoming the modern centre of 21st century Milan, Maison Moschino is the splendid result of a design project supervised by Rossella Jardini in cooperation with Jo Ann Tan.

By applying its fashion flair and design to the hotel industry, Moschino has created Maison Moschino, a stellar example of a new approach in hotel hospitality. The typical language of fashion has been transformed and adapted to create fanciful settings and surrealistic images for rooms in which fairytales, which tell an optimistic story about a fantasy world, come to three-dimensional life. "Alice's Room", "The Petal Room", "Red Riding Hood", "Forest", and "Gold" are the names of the rooms (16 different designs) that transform sleep into a dreamy experience, continuously weaving between a fairytale dream and upbeat, optimistic reality.

位于米兰Viale街Monte Grappe 12号的莫斯奇诺酒店被重新贴上了世界著名时尚品牌的标签。在酒店的名字前面加上"Maison"一字,设计师莫斯奇诺直接了当地表达出想将这家酒店转变成一个无与伦比的梦幻家园的想法,客人在这里休息将感受到前所未有的亲切与个性表现。

"当我进入酒店,感觉仿佛在家一般,所有的地方都似曾相似,温暖、亲切的氛围让你感受到安全感。我想将它按照原样保留,然而我知道,它就像一件装置,就本性而言是注定需要改变的。但是我相信改变也不会推翻它的根源。它与其他的酒店完全不同,它有勇气去展现自己原有的高品质工艺。这是一个如此特别的地方,这也就是为什么它叫做莫斯奇诺酒店。"当介绍莫斯奇诺酒店时,莫斯奇诺的创意总监罗塞拉·加蒂尼在展示她们最新的时尚酒店时这样说道。这里是家亦是被施了魔法的仙境,酒店拥有65间客房,一间由主厨莫雷诺·切德雷尼掌管的"隐秘米兰"餐厅,一间酒吧,一间"古缇"SPA以及莫斯奇诺的官方精品店。

酒店坐落于一座新古典建筑内,原是米兰第一座火车站,如今这里已经成为米兰21世纪的现代城市中心。酒店是由莫斯奇诺的创意总监罗塞拉·加蒂尼携手橱窗设计师乔·安·谭共同设计的杰作。

通过将时尚鉴赏力与设计元素投入酒店业,莫斯奇诺创作出的莫斯奇诺酒店可谓开辟了一条设计酒店的新道路。将典型的时尚语言改变得更适用于酒店设计,创作出梦幻的酒店场景布置和超现实主义的客房,在这里有童话,在这里讲述的是一个梦幻世界的乐观故事,讲述的是一个美梦成真的3D生活。客房被命名为"爱丽丝的房间"、"花瓣房间"、"小红帽"、"森林"和"金色",不同的客房名称将普通的睡眠变成了一场梦幻的体验,客人在童话般的梦境及乐观向上的现实之间自由转换。

| | |
|---|---|
| 1. Guestroom | 1. 客房 |
| 2. Bathroom | 2. 浴室 |
| 3. Corridor | 3. 走廊 |
| 4. Staircase | 4. 楼梯间 |

1. The surrealistic decoration in the lobby
2. The reception in the color of white
3. The reception decorated with interesting lamps and lights
4. Zzzzzzzzzzzzz room
5. A corner in Clouds room
6. The layout of Luxurious Attic room
7. The Forest room
8. The relax area in Art Spa
9. The Z background in Zzzzzzzzzzzzz room
10. The detail in The Forest
11. The detail in Luxurious Attic
12-14. The details of Ivy room
15. Tea table in Alice's Room
16. The corridors in the hotel
17. Blue room
18. The atrium decorated with clouds lights
19. The corridor in Art Spa
20-21. The chairs wearing ball gown in the restaurant
22. The clouds form lights in the guestroom
23. Sleeping in a Ballgown
24. The Ribbon room
25. Wallpaper room
26. Life is a Bed of Roses

1. 大堂内的超现实主义装饰品
2. 白色色调的接待区
3. 接待区装饰着有趣的落地灯和吊灯
4. Zzzzzzzzzzzzz主题客房
5. 云朵主题客房的一角
6. 奢华阁楼客房的布局
7. 森林主题客房
8. 艺术水疗中心的休息区
9. Zzzzzzzzzzzzz主题客房的Z字布景
10. 森林主题客房细节
11. 奢华阁楼客房细节
12-14. 常春藤主题客房细节
15. 爱丽丝的房间主题客房的茶杯茶几
16. 酒店走廊细节
17. 蓝色主题客房
18. 装饰着运动灯的中庭
19. 艺术水疗中心走廊
20、21. 餐厅中穿着晚礼服的座椅
22. 卧室内的云朵灯
23. "穿着晚礼服入眠"主题客房
24. 蝴蝶结主题客房
25. 壁纸主题客房
26. "生活就是拥有一床的玫瑰"主题客房

# "Unfinished" Aesthetic Values from the Genius of Deconstructivism

解构怪才的"未完成"美学

Maison Martin Margiela was set up in 1988 by Martin Margiela, a graduate of the Royal Academy of Fine Arts in Antwerp who worked as assistant to Jean-Paul Gaultier for three years. From 1998 to 2005 he was artistic director for the Hermès women's collections. Today Maison Martin Margiela has shops in France, Japan, Italy, the UK, the USA, Hong Kong, Germany, Dubai, Korea, Taiwan and of course Belgium.

Maison Martin Margiela is known for its predilection for white, for playing with contrasting proportions, for surrealism and for Pop culture. Maison Marin Margiela, precursor and pioneer, is also called the genius and master of deconstructivism. Heavily influenced by Rei Kawakubo — a Japanese pioneer designer whose design subverted the traditional fashion, Margiela is famous for deconstructing and reorganising clothing. He can easily see through the constitution and materials of clothing, and then deconstruct and transform them into unique pioneer fashion work. In 2009, Maison Martin Margiela was given carte blanche for the Elle Décoration suite, a 220m$^2$ apartment on the top floor of Palais de Chaillot. Maison Martin Margiela's identity expresses itself in parallel, and in the same place, through its fashion collections and interior design work. In the hotel, his Paris headquarters and the shops worldwide you see the white cotton covers, the trompe l'oeil, the subversion of objects and materials, the mixing of styles and eras, the play on aesthetic language and the humour as a permanent feature. Clothing, objects and interior design all communicate the same aesthetic values: a sense of detail, surrealism and lowbrow culture, oversizing and 2D projection, imaginatively recycled materials. all of these create an "unfinished" finish, and the rest part needs your sensibility and imagination to fullfill.

梅森·马丁·马吉拉品牌成立于1988年，由比利时著名的鬼才设计师马丁·马吉拉创建。马丁·马吉拉曾在著名的安特卫普皇家艺术学院学习，之后成为让－保罗·高缇耶的助手，为其工作过三年。在1998年到2005年这段时间，他成为爱马仕女装的创意总监。如今的梅森·马丁·马吉拉在世界各地都有店面，包括法国、日本、意大利、英国、美国、中国香港、德国、阿联酋迪拜、韩国、中国台湾，当然也包括比利时。

梅森·马丁·马吉拉以偏爱白色著名，擅长玩比例差异的游戏，设计倾向超现实主义和流行文化。梅森·马丁·马吉拉可以说是极简主义的先驱者和实践先锋，更有人称他为解构怪才，解构主义大师。深受日本先锋设计师川久保玲引起的颠覆传统时尚风潮的影响，马吉拉以解构及重组衣服技术而闻名，他能轻易的看出衣服的构造和布料特性，将其解构改造成一件独具个性的先锋时装。2009年，梅森·马丁·马吉拉开始将他的天赋施展向室内设计界，为夏约宫的艾丽装饰套房——一间220平方米的底层公寓进行室内设计。梅森·马丁·马吉拉在服装设计和室内设计中找到了共同的美学表现方法。在香榭丽舍家园酒店和品牌的时尚设计中都能看到用来覆盖的白布、拼贴画，对材料和物体的颠覆，时代和风格的融合，美学语言的运用和永恒的幽默感。品牌的理念认为，服装、饰品乃至室内设计都在讲述同一种美学价值：凭借一种对细节的感知，用超现实主义和通俗文化的精髓、扩张的二维影射以及充满想象力的对材料的循环利用，创作出一件仿佛是"未完成"的艺术作品，剩余的部分就要由观者各自的感受力和想象力来填补。

# La Maison Champs Elysées

香榭丽舍家园酒店

**Completion date:** 2011
**Location:** Paris, France
**Size:** 7000 m²
**Designer:** Maison Martin Margiela
**Photographer:** Martine Houghton

完成时间：2011
项目地点：法国，巴黎
规模：7000平方米
设计师：马丁·马吉拉设计公司
摄影师：马丁尼·霍顿

In 2010, Maison Martin Margiela, chosen by competitive tender, worked on - La Maison Champs Elysées - the older part of the Maison des Centraliens, dating from the days of Napoleon III. They redesigned the space to create suites, a restaurant, a smoking room, a bar and the reception hall. Maison Martin Margiela imagined this project as a direct continuation of its own artistic history, proposing a place with harmonious contrasts and a surrealistic slant. To carry out the project, Maison Martin Margiela worked jointly with other artists including a landscaper and some lighting designers.

Maison Martin Margiela wishes to offer a surprising poetic experience, one that cannot be exhausted in a single visit or a single stay. An experience of freedom, a journey within a journey, to be found nowhere else.

They are based on an offbeat take of standards, as symbolised at the outset by the paving in the reception hall, where black marble cabochons take liberties with the rule that says they must be placed at the corners of the white flagstones.

### Irony

Irony in the literal sense of the word, meaning the deliberate play on what is said as opposed to what is meant, letter as opposed to spirit, appearance as opposed to reality. The cabochons in the French-style paving are indeed there, but not in their rightful place. In the White Lounge, the spotlight rails illuminate only the traces of old picture frames – but these are painted onto new walls. In the guestrooms, the traditional Persian rugs are in fact patterns woven into the carpets. In the suites, the 19th-century mouldings are randomly interrupted. Playing with the vestiges of time in a new setting; a supremely dandified refinement suggestive of Beau Brummell who, it is said, had his clothes worn by his valet before donning them himself.

### Illusion

In the restaurant, the chairs and tables seem to be suspended a few centimetres above the floor but fear not, they are stable and comfortable. Trompe l'oeil reproduces the mouldings in the Empire reception rooms on the landings leading to them; lighting effects create the illusion that a closed door is open, allowing sunlight to filter in. In fact everything helps to create a theatrical world imbued with the magic of a show in which we are, if not actors, at least willing accomplices

### Respect

Respect for the building and the constraints of its heritage: walls or ceilings are not concealed unless for technical reasons. Maison Martin Margiela has not covered the mouldings or marble in the foyer to plaster one style over another. On the contrary, the intention was to further enhance the historic features of the place by dramatising them. Respect for the demands of comfort too, as demonstrated by the care taken with lighting, particularly in the bedrooms, and acoustics, especially in the restaurant. And, of course, respect for the project's inherent safety imperatives.

2010年,马丁·马吉拉设计公司肩负起一项富有竞争性的设计任务,翻新中央大厦这座历史可以追溯到拿破仑三世时期的建筑中的老酒店——香榭丽舍家园酒店。他们重新对空间进行了设计,以此创造出套房、餐厅、吸烟室、酒吧和接待大堂。马丁·马吉拉设计公司把他们的项目想象成对这座建筑的艺术史的一种直接延续,提议将空间设计的具有和谐的冲突感并富有超现实主义倾向。为了完成这个项目,马丁·马吉拉设计公司和其他艺术家共同合作,这其中还包括一位景观设计师和灯光设计师。

马丁·马吉拉希望提供一种令人惊喜的诗意般的体验,一种不会让单独的旅行或单独的停留感到筋疲力尽的体验。这是一种对自由的体验,一次旅行中的度假,在任何其他地方是感受不到的。

他们是基于在标准基础之上的标新立异,最具象征意义的是开始于接待大堂的地面铺设方式,这里用被磨圆的黑色大理石打破常规的地面铺设方法,并没有将这种凸型的圆石铺设在白色石板的四角。

### 讽刺

讽刺是一个文学味道十足的词,意思是估计将词语的本意反过来说,用词与原意相反,表现与现实相反。法式风格的地砖面上确实有被磨圆的大理石,但是却不在正确的地方。在白色的休息室,聚光灯柱照亮的仅是古典画作的边框——但是这些画是被直接画在墙上的。在卧室里,传统的波斯地毯事实上是和普通地毯编织在一起的。在套房内,19世纪的装饰线条在室内随意的出现,打断着客人的视线。在新的布景中玩味时间的痕迹,仿佛是一位衣着华丽的花花公子将他的衣服套在了管家的身上。

### 幻象

在餐厅,椅子和桌子似乎都是腾空高于地面几厘米的,但是却不用担心它们的稳定性和舒适度。在帝国接待室里,用法国新视觉画法复制出的装饰条纹仿佛在降落的过程中牵引着这些桌椅,灯光制造出幻觉,仿佛门是开着的,让阳光透了进来。事实上,所有的一切协助打造出一个戏剧般的世界,客人们在这场秀中仿佛被施了魔法,如果不是这里的演员,也至少是参与者。

### 尊敬

对于建筑和遗产的尊敬包括:墙或者是天花板被隐藏了起来,除非是技术上的原因不能达到的。马丁·马吉拉设计公司没有遮住旧建筑的装饰条纹或者是门厅的大理石,以此用来减弱一种风格对另一种风格的影响。相反的,设计的目的是通过将这些细节戏剧化来进一步加强这个地方的历史特色。尊重还体现在对舒适度的要求上,可以证实的其中一点是体现在灯光上,特别是在卧室里,此外对音响效果的特别设计在餐厅设计中尤为突出。当然,项目安全性的要求也必须是要受到重视和尊重的。

1. The entrance to La Table du 8 Restaurant
2. The living room in The Gilded Lounge Suite
3. A corner to bathroom in The Gilded Lounge Suite
4. A corner in The White Lounge
5. A corner in The Gilded Lounge Suite
6. The bedroom in Curiosity Case Suite
7. The customized chair in Curiosity Case Suite
8-9. The corner in Curiosity Case Suite
10. The black corridor to The White Lounge
11. The detail in corridor
12. The door of The White Lounge
13. The lobby with creative French style

1. 8号桌子餐厅入口
2. 镀金沙龙套房客厅
3. 镀金沙龙套房通往浴室的一角
4. 白色沙龙一角
5. 镀金沙龙一角
6. 珍奇箱子套房卧室
7. 套珍奇箱子房内的定制座椅
8、9. 珍奇箱子一角
10. 通往白色沙龙的黑色走廊
11. 走廊细节
12. 白色沙龙的门
13. 创意法式大堂

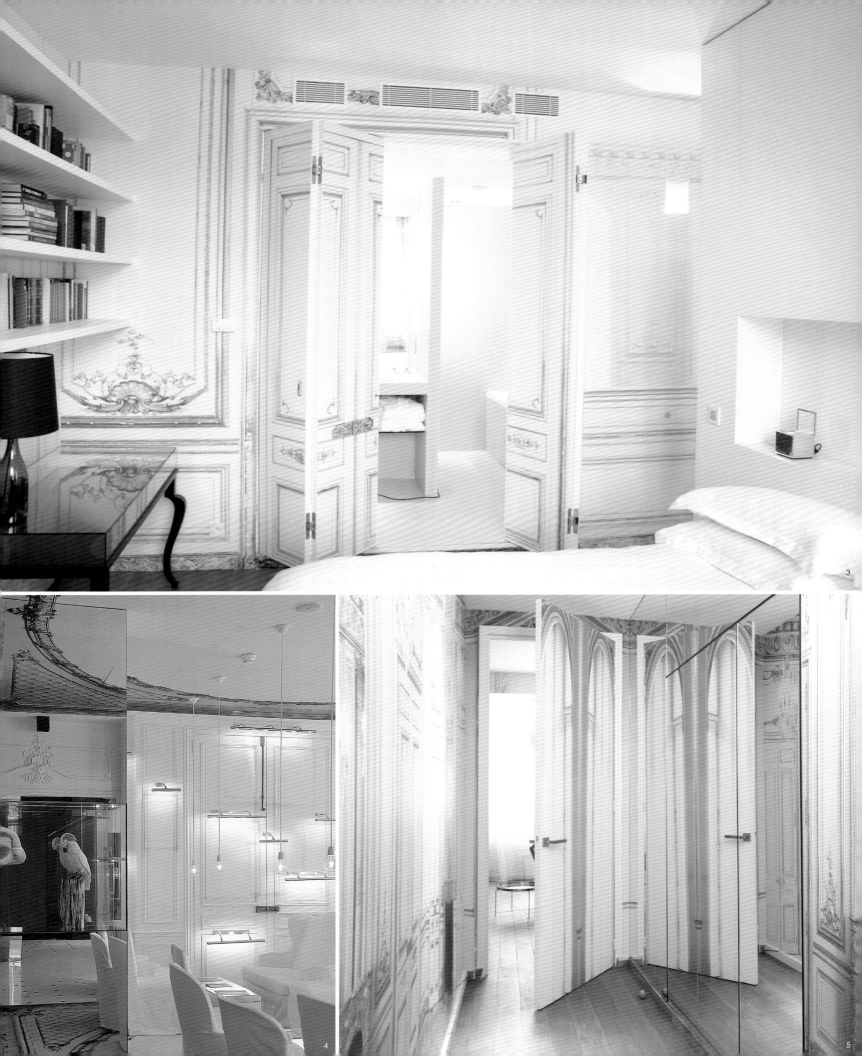

# A Story of Enchanted Hotel Comes True

酒店故事，魔幻成真

Stella Cadente had studied fashion design at Studio Berçot and at FIT in New York. After graduating Stella sharpened her skills by working for the leading Parisian couture houses. In 1995 she created her own ready-to-wear brand and opened her first boutique. Stella Cadente creates like we tell the stories that are sometimes light, sometimes mysterious, but always with an element of fantasy and never without spirit. The designer often founds her inspiration in the tales whose dark and disturbing side enthralls her particularly. She also likes to capture the current atmosphere and the street fashion trends. The fantasy world she created, always tinged with a touch of rock"n"roll, has become a substantial part of the codes of her brand.

When she heard for the first time about the renovation of the former Hotel Lyon-Mulhouse, Stella Cadente considered it as an extraordinary opportunity for the creative expression. Designer has embodied her ideas in the all 38 rooms and public areas of this hotel. The great relationships and trust between the hotel managers and Stella Cadente, allowed to fulfill the initial ideas without distorting them. Most of the furniture pieces and objects that currently decorate the Hotel Original, have been specifically designed by Stella Cadente. Moreover these decorative pieces are strikingly similar to the very early designer's creations. According to the expression: "she dresses up the houses" like she dresses up human beings. The creative impulse is the same no matter what the field is. Stella Cadente's creations are like stories.

Meanwhile, endowed with the plots of several stories, the hotel is vitalised. There is the throne of Ice Queen, the colourful forests in the enchanted world, the princess wearing donkey skin and Lewis Carroll's poetic world. A complete creator, she finds her inspiration in fairy tales, poetry, and street trends from which she selects and transforms both fantasy and darkness as she wishes. The Hotel ORiginal presents itself as an range of her unlimited imagination, as a perfect reflection of her enchanting and dynamic world!

斯黛拉·卡丹特曾在法国高等服装设计学院以及纽约时尚理工学院学习时尚设计。在毕业之后斯黛拉在巴黎顶尖的高级成衣工作室工作，使得她的技艺得到磨练。1995年，她创立了自己的成衣品牌，她的第一家精品店也随之开业。斯黛拉·卡丹特创造了传说，她的作品有时是轻柔的，有时是神秘的，但却总是略带些梦幻元素，从不缺灵气。她的设计灵感来自于那些传说中的黑暗和令人不安的一面，这些元素特别令她着迷。她还喜欢去捕捉流行气息和街头时尚。她创造出的梦幻世界总是渲染了一层摇滚色彩，这已经成为品牌潜在的定式。

当她开始着手设计这家前身为里昂·米卢斯的酒店，斯黛拉·卡丹特想把此作为一次非比寻常的机会，以此用来表现她的创造力。在酒店的一共38间客房和公共区域内，到处都表现出她在时装中难以全面表达的理念。酒店经理和斯黛拉·卡丹特之间的友好关系和对彼此的信任，让设计师最初的想法得到完整的实现，而没有受到任何外部的干扰。大部分的家具和酒店的装饰物都由斯黛拉·卡丹特特别设计。此外，这些装饰作品与设计师之前的服装设计作品有很大的相似性。"她为房子穿衣打扮"就如同她给人装扮一样。无论是哪个领域都有着同样的创作动力。斯黛拉·卡丹特的作品就像一个个故事，而酒店中同样是被赋予情节的一个个故事，一个个被赋予生命的空间，这里有冰雪皇后的宝座，魔幻世界的彩色深林，驴皮公主的倩影和刘易斯·卡罗尔的诗意世界。这样的酒店印证了卡丹特是一个完全的创作者，她将那些从神话、诗歌得来的灵感去粗取精，按照她的愿望转变成同时具有梦幻色彩以及黑暗元素的作品。原创酒店表达了设计师一系列无限的想象，是她魔幻且充满活力的时尚世界的完美再现。

# Hôtel ORiginal

原创酒店

**Completion date:** 2012
**Location:** Paris, France
**Size:** 38 rooms
**Designer:** Stella Cadente
**Photographer:** Christophe Bielsa

完成时间：2012
项目地点：法国，巴黎
规模：38间客房
设计师：斯黛拉·卡丹特
摄影师：克里斯托弗·贝尔萨

Next to both the lively Place de la Bastille and the very chic Place des Vosges, the Hotel ORiginal is ideally located and offers its guests multiple ways to enjoy Paris. Whether through fashion, the world of perfumes or decorative arts, Stella Cadente expresses her creativity with great freedom and enthusiasm.

According to the expression: "she dresses up the houses" like she dresses up human beings. The creative impulse is the same no matter what the field is. Stella Cadente's creations are like stories. A complete creator, she finds her inspiration in fairy tales, poetry, and street trends from which she selects and transforms both fantasy and darkness as she wishes. The Hotel ORiginal presents itself as an range of her unlimited imagination, as a perfect reflection of her enchanting and dynamic world!

### * THE LOUNGE

The lounge, with its subdued, warm colours, was created as a comforting cocoon, and is the perfect place to enjoy breakfast or drinks, in a unique atmosphere. Materials, furniture, plays of light and original design inhabit the space, which becomes the threshold to the guest's enchanted journey at the Hotel ORiginal.

### * CRYSTAL QUEEN

This room was inspired by Andersen's fairy tale. It plays with its paradoxical interior design: it is pure and cosy, just like a precious ice cocoon. While the rhinestones and crystals which sparkle under the subtle rays of light warm up this immaculate nest, the fur, the majestic bed-head, and the throne-like baroque armchair remind you that you're entering the den of a Queen…

### * JOKER

Playing on both universal and mysterious graphic characteristics, this room is a true gateway to a chic and playful world. Black, Red, and White contrast with one another with force and elegance. Waxed surfaces and large fabrics soften the lines and checkered patterns which cover this space, watched over by the enigmatic female Joker. She reveals herself on the ceiling under a streak of dark light…

### * THE 7 DEADLY SINS

The seven rooms on the last floor are smaller and have mansard windows. They have been designed as a tribute to the seven deadly sins, staged in a mix of golden, bronze highlights and brown shades. Each den displays an original photograph by Stella&Claudel, celebrating subversion through subdued and comforting lighting. The thick carpet and the fur quilt enhance how comfortable these forbidden nests are.

### * BLUE ENCHANTED FOREST

This room associates imaginary and real elements related to the forest, a theme the designer adores. It is like a journey into the woods, a fantasy world where imaginary creatures and authentic creations taken from a comfortable country house meet. Bright colours, gilt, shiny mosaic and patterns of light and shadow turn this contrasted space into a reassuring and mysterious place, to be discovered in the moonlight…

### * YELLOW ENCHANTED FOREST

The world of the Enchanted Forest is a boundless source of inspiration for the designer. This room is clad in warm shades of bright yellow. It is the mirror image of the Blue room, and it displays a dreamlike world in which lamps turn into fantasy creatures, golden frames become alive and leaf patterns are revealed in your bed at sunset…

### * WONDERLAND

This room was inspired by Lewis Carroll's poetic world, and it transcribes the double facet of Alice's Wonderland through a combination of light purple, striped with black, and of subtle patches of red, as many references to the Queen of Hearts. The checkered board, the Rabbit and the Cheshire Cat finalise this ambivalent setting, which associates classical and fantasy elements…

### * GOLD

The designer was inspired by her memories of Donkey Skin's moon-coloured and sun-coloured dresses to create monochromatic rooms with intense highlights. The Gold Room is like a precious case in which some walls have been covered with gold foils enhanced by a unique set of lights. While the ubiquity of gold may evoke stones from Haussmann buildings, it may also be seen as a luxurious cocoon, ideal for daydreaming.

### * SILVER

The other monochromatic room inspired by Donkey Skin's moon-coloured and sun-coloured outfits offers as welcoming an atmosphere thanks to various shades of silver. The combination of materials, symbols, and lights gives birth to a soft and purified world, which allows the guest's imagination to wander freely…

1. The 7 Deadly Sins room on the top floor
2. A corner with mansard windows in The 7 Deadly Sins
3. Blue Enchanted Forest Room
4. A colourful shiny mosaic wall in Blue Enchanted Forest room
5. A view from the mirror in Blue Enchanted Forest room
6. The bathroom in Joker room
7. The layout of Joker room
8. The layout of Silver room
9–10. The detail of Silver room

1. 顶楼的七宗罪主题客房
2. 七宗罪的斜窗一角
3. 蓝色魔幻森林主题客房
4. 蓝色魔幻森林主题客房的闪耀绚丽的马赛克墙
5. 镜子里的蓝色魔幻森林主题客房
6. 小丑主题客房的浴室
7. 小丑主题客房布局
8. 银色主题客房布局
9、10. 银色主题客房细节

原创酒店比邻巴士底狱和孚日广场,坐拥理想的地理位置,让客人能够全方位的享受巴黎生活。无论是在时尚、香水和装饰艺术品的世界,斯黛拉·卡丹特用她的非凡率性和热情表达出创造力。

"她为房子穿衣打扮"就如同她给人装扮一样。无论是哪个领域都有着同样的创作动力。斯黛拉·卡丹特的作品就想一个故事。这是一个完全的创作者,她的灵感来自那些神话、诗歌和街头时尚,她将此去粗取精,按照她的愿望转变成同时具有梦幻色彩和黑暗面的作品。原创酒店表达了设计师一系列无限的想象,是她魔幻且充满活力世界的完美再现。

### 大堂休息区

大堂休息区采用柔和温暖的色调,这里被设计成一个舒适的茧,一个具有独特氛围,可以享受早餐和饮品的完美场所。各色材料、家具、灯光的巧妙运用和原创的设计栖居在这个空间,开启了客人们的奇幻之旅。

### 水晶皇后客房

这个房间受到安徒生的童话故事的启发。室内设计充满矛盾色彩:纯洁而惬意,像似珍贵的冰之屋。人造石和水晶在光线的照耀下光芒四射让整个空间温暖起来,毛皮、豪华的床头,还有宝座一样的巴洛克式扶手椅提醒着客人,这里是皇后的寝宫。

### 小丑客房

装饰着宇宙图案和神秘图案,这间客房仿佛是通往潮流游乐世界的真实通道。黑色、红色和白色,强烈又优雅的在彼此之间碰撞、冲突。打过蜡的墙面和巨幅的织物使空间的线条变得柔,格纹的墙纸贴满了整个空间,这些都被天花板上的一个神秘女性小丑形象监视着。安置在天花板下的条状灯让这个小丑图象显示出来。

### 七宗罪客房

这七间客房在酒店的顶层,它们比其他客房略小,都有着复折屋顶窗。这些客房的设计都与七宗罪的主题相关,用金色和青铜色点睛,用棕色制造阴影。每一个房间都展示了一幅由斯黛拉和克劳戴尔摄影工作室拍摄的图片,通过柔和、舒适的灯光进行渲染。厚厚的地毯和毛皮被加强了这些禁忌巢穴的舒适性。

### 蓝色魔幻森林客房

这间客房将虚幻又真实的、与森林有关的元素联合起来,这是设计师钟爱的主题。仿佛是进入了丛林和梦幻世界的一次旅行,在这里虚幻的生物和仿佛是来自一间舒适乡村小屋的创作元素相互衬托。明亮的色彩、镀金、闪耀的马赛克,还有各式各样的光影效果让这个冲突的空间变成了一个令人安心的神秘之地,在月光中等待发现……

### 黄色魔幻森林

魔幻森林的世界是设计师取之不尽的灵感源泉。客房被亮黄色制造出的暖色影子覆盖。它是蓝色客房的镜面图像,在这里上演的是梦幻世界的朝朝暮暮,落地灯成为了梦幻的生物,在黄昏,开启它时金属的画框和床头的树叶装饰立刻有了生命力……

### 仙境客房

这间客房的设计灵感来自刘易斯·卡罗尔的诗意世界,它将爱丽丝的双面仙境用紫色的灯光、黑色的条纹、红色的皇后桃心图案再现了出来。条格墙板,兔子和喵喵猫的形象使这个将古典元素和梦幻元素的矛盾场景完整了起来……

### 金色客房

设计师的灵感来自于《驴皮公主》的故事中月亮色和太阳色的礼服。金色客房仿佛是一个珍贵的首饰盒,在这里墙面用金箔覆盖并用特有的一组灯光加强效果。客房主导的金色让人联想起奥斯曼时期的建筑,奢华感至极,仿佛白天也处在梦境。

### 银色客房

另一间单色客房同样受到《驴皮公主》故事中月亮色和太阳色的礼服的启发,通过各种的银色营造出一种热情的氛围。通过各种材料、符号和灯光的结合,一个温柔、纯净的世界诞生了,让在这里的客人们任由想象力的漫游……

# Fashion Brands Hotels
## 时尚品牌的酒店业入侵

The charm of fashion brands hotels is that they make guests surrounded by famous brands they love. The most direct way is to inject the existing brand style and elements into the process of the hotel interior design. Many designers advocate that fashion crossover design hotels can effectively convey the philosophy of brands at a broad level. Moreover, designers convert their demands into a lifestyle to influence people. In Missoni hotels, classic round-dot prints are noticed from time to time; in Palazzo Versace, antique feeling of luxury is the faithful representation of Versace fresh and colourful style; and the most well-known Armani Hotel Dubai also extends its simplicity and elegance; The eclecticism always in Tcherassi is also infused into Tcherassi Hotel + Spa, with ballet - like elegance in the colonial architecture. However, Ferragamo is not of the same kind. Unlike other fashion hotels, its brand logo cannot be found and the word "Ferragamo" is not mentioned on its billboard either. Instead of despising advertising and promotion, Ferragamo does not regard brand as transient business brand and what it considers most is to reinforce its eternal classic image in consumers' mind.

让旅行沉浸在自己喜爱的时尚品牌之中，这是时尚品牌酒店的吸引力所在。这类酒店将品牌风格和元素融入到酒店的室内设计中。很多设计者认为这类酒店能够更宽而有效的传达品牌哲学，他们想将自己的追求转变成一种生活方式去影响大众。在米索尼的品牌酒店中，客人会在酒店不时的看到米索尼的经典圆点印花；在范思哲宫廷酒店，奢华的古典气息正是范思哲服装展现的华美风格；最为人所熟知的阿玛尼迪拜酒店同样延续着品牌服装的极简优雅；切拉西品牌一贯的折衷主义也体现在品牌酒店中，硬朗的殖民建筑中有如芭蕾舞般优雅的温柔氛围；菲拉格慕可能要算上这类酒店的一个另类，不像其他时尚品牌酒店，他们的酒店并没有明显的打上品牌的标签，连酒店名称也不提菲拉格慕一个字，这并非菲拉格慕不重视品牌的理念宣传，相反认为品牌并非追求潮流变迁商业品牌，他们只重视在酒店的设计之中融入菲拉慕格沉稳和永恒的经典印象。

# The Hotel Design Dream of the Giant in Fashion Industry

时装界巨子的酒店设计之梦

George Armani, who is a world famous Italian fashion designer, was born in Piacenza Italy. He had studied medicine and photography and had been a men's clothing designer in Charity. He set up his own George Armani company in 1975. He's got the Naimen Marcos Prize, Wool Mark Award, Life Achievement Award, the United States International Designer Association Award, Kuti Sackler Award, and so on. George Armani is now a European designer's brand that has the biggest sales in America. It may be said that Armani is the finest example in the fashion world which combine the fashion art with commercial creation to its best. And its high profit in the industry always leaves the other brand too far behind. Of course this fashion giant won't put his design spirit just into clothing design.

The design style of Armani is neither fashionable nor traditional. It can make an almost perfect and amazing balance between the market demand and elegant fashion. His design and tailor are elegant, which get rid of all the irrelevant details. This kind of simple breaks the bounds between tradition and reality, masculine and tender. It is reflected in the decoration style of Armani Hotel. There's no artwork hanging on the walls of the guestrooms. According to the staff's saying, it is for keeping the "simpleness and gracefulness" of the hotel. The magnificent guestrooms and suites in Armani Hotel Dubai present the simple luxury with each detail reflects the Armani style. Combined with rare materials and high style facings, the practical and elegant design concept of Armani creates an indoor environment which shows the personal charm. It took about 5 years to construct the hotel. Giorgio Armani inspected almost all the design in person, including Eramosa stone floor, zebra wood screen and wallcloth, as well as the accessories in bathroom and spa centre, etc. It realised Giorgio Armani's dream for many years to achieve this only high quality hotel.

乔治·阿玛尼，这位举世闻名的意大利时装设计师出生于意大利皮亚琴察，学习过医药及摄影，曾在切瑞蒂任男装设计师，1975年创立乔治·阿玛尼公司。曾获奈门－马科斯奖、全羊毛标志奖、生活成就奖、美国国际设计师协会奖、库蒂·沙克奖等奖项。乔治·阿玛尼现在已是在美国销量最大的欧洲时尚品牌，在时尚界阿玛尼可谓是将时尚艺术与商业创意连接的最恰到好处的典范，经营上的高利润让其他品牌一直望尘莫及。这样的一位时尚界巨子当然不会只将他的设计灵感投入单一的成衣设计。

阿玛尼以使用新型面料及优良的制作而闻名。他的设计风格既不潮流亦非传统。它能够在市场需求和优雅时尚之间创造出一种近乎完美、令人惊叹的平衡。他的设计裁剪优雅，无关的细节全部去掉。这种简约打破了传统与现实、阳刚与温柔的界限。阿玛尼酒店的装修风格体现了这一点。客房墙上没有悬挂画作，按酒店工作人员的说法，这是为保持酒店的"简约和优雅"。阿玛尼迪拜酒店的华丽客房和套房演绎简约奢华，每个细节也投射阿玛尼风格。实用而高雅的阿玛尼设计信念，配合名贵选料和格调高尚的饰面，创出饶显个人风韵的室内环境。酒店耗时近五年建造，乔治·阿玛尼全程亲自监督所有设计，包括加拿大木纹石地板、斑马木饰屏和墙布，以及浴室和水疗中心配件等，成就只此一家的臻品酒店，一圆乔治·阿玛尼多年的梦想。

# Armani Hotel Dubai

迪拜阿玛尼酒店

**Completion date:** 2010
**Location:** Dubai, UAE
**Designer:** Giorgio Armani
**Size:** 40,000 m²
**Photographer:** Armani Hotel Dubai

完成时间：2010年
项目地点：阿联酋，迪拜
设计师：乔治·阿玛尼
规模：40,000平方米
摄影师：图片由阿玛尼迪拜酒店提供

Soaring high above Downtown Dubai in the iconic Burj Khalifa, the world's tallest tower, Armani Hotel Dubai is the world's first hotel designed and developed by Giorgio Armani. Reflecting the pure elegance, simplicity and sophisticated comfort that define Armani's signature style, the hotel is the realisation of the designer's long-held dream to bring his sophisticated style to life in the most complete way and offer his customers a Stay with Armani experience. Every detail in the hotel bears the Armani signature, beginning with the warm Italian-style hospitality and moving to each element of the design from the Eramosa stone floors to the zebrawood panels, bespoke furnishings and personally designed hotel amenities.

Sophisticated colours, clean lines and unique textures blend together seamlessly with the tower's stunning architecture and natural light to create an atmosphere of calm serenity where guests can retreat into a world of minimalist elegance.

Armani Hotel features eight restaurants offering a wide choice of world cuisines ranging from Japanese and Indian to Mediterranean and authentic fine-dining Italian. The impressive culinary options reflect the cosmopolitan nature of the city.

An oasis of peace and tranquility in the heart of a bustling city, the 12,000 square feet Armani/SPA reflects the Armani lifestyle and design philosophies, offering beautiful unique spaces and outstanding service for individually personalised treatments, personal fitness, sequential thermal bathing, creative spa cuisine, or simply private and social relaxation.

Every guest at the Armani/SPA receives a personal consultation from the spa manager to develop a bespoke sensory experience designed by Armani. The spa therapies have been designed to fulfill different goals. MU quenches a desire for relaxation and stillness; LIBERTA' encourages freedom of movement and the release of physical pain; and FLUIDITA' enhances vitality, restoring internal balance.

The hotel offers a luxurious ballroom seating up to 450 people; a stunning outdoor pavilion area overlooking The Dubai Fountain; a charming Majlis offering an Arabic style meeting room, and several meeting lounges and boardrooms.

From exquisite centrepieces by Armani/Fiori to bespoke Armani/Dolci gifts, every detail is thoughtfully put together with the help of the expert planners. Harmonious understated décor, carefully selected tableware and linens, perfected menus and impeccable service supported by state-of-the-art technology ensure guests experience the Armani lifestyle promise.

在高耸入云的迪拜世界第一高塔——哈利法塔内，由Giorgio Armani亲自设计的阿玛尼迪拜酒店就坐落在这里。阿玛尼迪拜酒店是第一家由乔治·阿玛尼设计并开发的酒店。酒店所反映出的纯粹的优雅，朴素以及精致的舒适感正是阿玛尼品牌的标志风格。这座酒店的落成是设计师本人将自己长期以来的梦想融入到品位生活中的一种最全面的展示，客人在这里体验到的是一次阿玛尼之旅。酒店的每一个设计细节都彰显阿玛尼特色，进入酒店就被浓厚的意大利式温暖氛围包围，随之而来每个设计元素包括加拿大花纹地板、斑马木隔板、定制家具和量身定制的酒店设施都散发着阿玛尼式的品位。

精致的色彩，清晰的线条，独特的材质互相糅合，再与酒店塔楼的惊世建筑与自然光照融合在一起营造出的是一种平静安宁的氛围，在这里，客人可以逃离世俗，进入一个极简主义的高雅世界。

阿玛尼酒店内设有8家餐厅，提供全世界范围内的各类美食，从日式料理，印度大餐，到地中海美味，纯正的意大利美食。这些让人印象深刻的美食餐点正反应出这座城市的大都市本质。

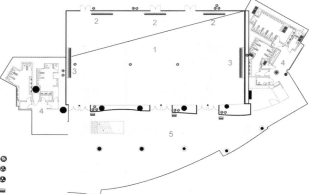

1. Armani ballroom
2. Small screen
3. Main screen
4. Toilet
5. Ballroom prefunction

1. 阿玛尼宴会厅
2. 小屏幕
3. 主屏幕
4. 洗手间
5. 宴会前厅

在这座喧嚣城市的中心是一座和平安宁的绿洲，12,000平方英尺的阿玛尼水疗中心反映出阿玛尼式的生活方式和设计哲学，这里提供美丽独特的空间和与众不同的服务，包括个人理疗服务，人健身专家、一系列热水洗浴设施、创意水疗餐或者私人简餐，以及社交休息室。

每一位在阿玛尼水疗中心的客人都会享受到私人水疗经理的专业咨询，为其定制有阿玛尼专业设计的感官理疗体验。专门设计的水疗处方针对不同的理疗目标。MU疗程满足客人放松静修的需要；LIBERTA'倡导自由运动，可以帮助客人舒缓身体病痛；以及FLUIDITA'加强身体活力，修复身体内部的平衡。

酒店提供奢华的宴会厅可以供450人共同用餐；令人惊叹的户外凉亭区遥望迪拜喷泉；迷人的吉利斯议会室提供给客人阿拉伯式的会议空间，包括多个会议休息室和董事会议室。

从阿玛尼花店Fiori里精美绝伦的装饰品到酒店的甜品店Dolci内定制的小礼物，每一处细节都至始至终贯穿着专业规划者的心血。和谐质朴的装饰，精心挑选的餐具和床用织物，完美无缺的餐点和有高端技术支持、无可挑剔的服务，这些都满足了客人在这里体验阿玛尼式生活方式的愿望。

1. Burj Khalifa in downtown Dubai
2. The entrance of Armani Hotel Dubai
3. The Japanese restaurant — Armani Hashi
4. The reception in the lobby
5. The lobby in Armani Galleria
6. The dining area in Armani Amal
7. The dining area in Armani Peck
8. A corner in Armani Prive
9. The corridor in Armani Hotel Dubai
10. Enoteca in Armani Ristorante
11. Private dining area in Armani Amal
12. Business centre
13. The living room in Armani Ambassador Suite
14. Foot bath ritual in Armani Spa
15. The bathroom of signature suite in Armani Spa
16. Bathwood shower
17. The living room in Armani Fountain Suite
18. The living room in Armani Premiere Suite
19. Armani Fountain Suite

1. 位于迪拜市中心的哈利法塔
2. 迪拜阿玛尼酒店入口
3. 日式料理餐厅——哈什餐厅
4. 大堂内的接待区
5. 阿玛尼画廊大堂
6. 阿玛尼阿迈勒餐厅就餐区
7. 阿玛尼派克餐厅的就餐区
8. 阿玛尼私密餐厅的一角
9. 酒店走廊
10. 阿玛尼意大利餐厅的安娜塔卡包房
11. 阿玛尼派克餐厅的私人就餐区
12. 商务中心
13. 阿玛尼大使套房的客厅
14. 阿玛尼水疗中心的足浴盆
15. 阿玛尼水疗中心标志套房内的浴室
16. 浴室淋浴
17. 阿玛尼喷泉套房的客厅
18. 阿玛尼优选套房的客厅
19. 阿玛尼喷泉套房

# The Brilliant Rebirth of Colours and Patterns

色彩与图案的舞动新生

Missoni is a fashion brand opened up by an Italian couple Ottavio Missoni and Portia Missoni in half a century ago. This brand is famous for its knitting shirts which are full of arts. After its development of half a century, now this brand has formed its own Missoni classic style in the fashion world with abstract geometric patterns, changeable lines group, and bright colours. Many celebrities such as Jacqueline Kennedy, Woody Allen are all very fond of it. It is not too exaggerated to call it the pronoun of knitting shirts in fashion world. Maybe there's no saying of stop for fashion. When she was 73 years old, Portia Missoni created a legend in the name of fashion. She delivered her fashion business to her daughter and diverted her concentration on the Missoni Home design. Whereby, the stripes, patterns and colours on Missoni knitting were painted on the fabrics and cup trays. Colours which may be the basic elements to Missoni design are used in the Missoni Home products. To print and dye different kinds of colours in space, which is called "Space-dye" by the designer herself, can make the space colour clean and transparent with advanced dyeing technology to use bright colours instead of pure colours. Then, the second feature of Missoni design is the geometric patterns and lines. In home products design, Missoni extends the lines to wavy lines, gridirons, round dots, and match them with Missoni's clean and bright colours through different kinds of combination, which makes the space dance fashionably. When Portia Missoni designed Missoni Edinburgh Hotel, these Missoni home products were endowed with a more realistic significance as if that this is a Missoni experience place, where the colours and patterns respected by the guests are really come to life.

米索尼（Missoni）是由一对意大利夫妇奥塔维袄·米索尼和罗西塔·米索尼在半个世纪前开创的时尚品牌。品牌以极富艺术感的针织衫而闻名，经过半个世纪的发展，目前品牌已经在时尚界形成了别具一格的米索尼经典风格，抽象的几何形图案，多变的线条组合以及亮丽的色彩。许多名人如杰奎琳·肯尼迪、伍迪·艾伦都对它喜爱有加，称它为时尚界的针织代名词毫不为过。也许时尚远没有止步一说，罗西塔·米索尼在她73岁时再一次以时尚之名创造了一个传奇，她将自己的时装事业转交给自己的女儿，自己改投专心设计米索尼家居产品线（Missoni Home），借此她将米索尼针织衫上的条纹，图案及色彩绘在了织物上、杯盘上。色彩对于米索尼的设计可谓是最基本的元素，用在米索尼的家居产品上，被设计师本人称为"Space-dye"，意思是用多样的色彩印染在空间之上，用先进的染织技术将绚丽的色彩取代纯色，使空间色彩干净、透明。其次，米索尼的第二个特色是几何形的图案和线条，在家居用品的设计上，米索尼将线条延伸成波浪线、方格、圆点，再经过各种组合搭配米索尼干净的绚色，让整个空间时尚地舞蹈起来。罗西塔·米索尼在设计爱丁堡米索尼酒店之时正是把这些米索尼的家居用品赋予了更现实的意义，在这里，仿佛是一个米索尼的体验场所，客人们崇敬的色彩与图案，被再次赋予新的生命力。

# Hotel Missoni Edinburgh

爱丁堡米索尼酒店

**Completion date:** 2009
**Location:** Edinburgh, UK
**Designer:** Rosita Missoni, Matteo Thun & Partners
**Area:** 8630 m²
**Photographer:** Beppe Raso; Paolo Riolzi

完成时间：2009年
项目地点：英国，爱丁堡
设计师：罗西塔·米索尼，玛窦·图恩建筑与室内设计事务所
规模：8630平方米
摄影师：柏普·拉苏，保罗·里欧兹

Opened in June 2009, Hotel Missoni Edinburgh was the debut hotel from The Carlson Rezidor Hotel Group and luxury fashion house, Missoni. Combining the unique style of the iconic Italian fashion & interiors house with the expertise of The Carlson Rezidor Hotel Group, Hotel Missoni redefines the design hotel. Designed to give guests a true taste of the Missoni lifestyle, Hotel Missoni has fast become one of Edinburgh's most desirable destinations, winning a number of awards and accolades since opening.

Opened in June 2009, Hotel Missoni Edinburgh was the debut hotel from The Carlson Rezidor Hotel Group and luxury fashion house, Missoni. Combining the unique style of the iconic Italian fashion & interiors house with the expertise of The Carlson Rezidor Hotel Group, Hotel Missoni redefines the design hotel. Designed to give guests a true taste of the Missoni lifestyle, Hotel Missoni has fast become one of Edinburgh's most desirable destinations, winning a number of awards and accolades since opening.

Each Hotel Missoni is reflective of its location and as a tribute to the Scottish heritage, Rosita has included pieces by Charles Rennie Mackintosh including his famous Ladderback Chair.

The lobby of Hotel Missoni Edinburgh also includes two large mosaic urns, which has become something of a Hotel Missoni trademark.

Hotel Missoni Edinburgh offers 136 rooms and suites, some of which enjoy stunning views over the city landmarks and the famous Royal Mile. The rooms continue the palette of black & white seen throughout the property and are animated with bursts of colour. Combining form and function, the accommodation has been carefully designed to meet the needs of today's modern traveller with linens chosen from the Missoni Home range.

In addition to the guest rooms, there are seven suites ranging in size from 44 sq metres to the 67sq metre, Suite d'Argento.

An award-winning destination restaurant in Edinburgh, Cucina lies at the very heart of the hotel. Awarded Best Italian Restaurant of the Year in 2011 and 2012 at the Scottish Restaurant Awards, and achieving a host of other awards since opening, Cucina has firmly established itself on Edinburgh's dining scene.

Blending fashion and function, Cucina is a vibrant, bustling restaurant and the decor displays the iconic brand throughout, from the fabulous Missoni patterned tableware to the black and white signature crockery.

Inspired by the Italian tradition of 'passeggiata' Bar Missoni is the place in Edinburgh to see and be seen. Positioned in the lobby of the hotel on the ground floor, Bar Missoni can host up to 50 people and serves a cocktail list that includes classic drinks, non-alcoholic cocktails and recipes created specially for Bar Missoni.

There are three unique and vibrant private rooms available for meetings or events, each stylishly designed in keeping with the Missoni interiors. The three rooms vary

in size from 28 sq metres to 75 sq metres and the largest room can be divided into two adjoining spaces. A fourth space (100 sq metres) is available for use as a breakout area – the perfect place for receptions and refreshments.

Situated on the second floor, Spa Missoni offers the latest in designer treatments to soothe, restore and replenish. Adorned in the signature Missoni patterns, the space challenges conventional spa décor, yet manages to create a luxurious and relaxing atmosphere. There are over 50 treatments for both men and women using exclusive brands Aromatherapy Associates, Eve Lom and Natura Bissé. Prices for treatments start from £40.

爱丁堡米索尼酒店开业于2009年，是雷兹达酒店集团和奢侈时尚品牌米索尼合作的第一家酒店。结合独特的殿堂级意大利时尚品牌的时尚风格和卡尔森雷兹达酒店集团管理酒店的专长，米索尼酒店重新定义了设计酒店。酒店带给客人一种真实的米索尼生活方式，这让酒店快速的成为爱丁堡最让人期待的目的地之一，从开业以来就获奖无数。

酒店由米索尼创意总监罗西塔·米索尼亲自设计室内，整体设计大胆、直率且富有戏剧性。酒店的整体配色方案以简单为主，时尚的黑色与白色之中不时闪烁着其他艳丽的色彩。罗西塔用清晰的米索尼印花搭配一些她喜爱的艺术家作品，例如马塞尔·万德斯福、埃罗·沙里宁和阿恩·雅各布森的作品。客人会发现这些大师的作品遍布在酒店的7个楼层，其中一些是罗西塔的个人收藏品，例如汉斯·J·韦格纳的如愿骨椅子。

米索尼酒店反映了酒店所在城市的特色也为苏格兰遗产保护做出了贡献。罗西塔在酒店中收录了麦金托什的作品，包括著名的梯背椅。

在大堂，酒店还收录了两件马赛克陶瓮，这已经成为了米索尼酒店的标志。

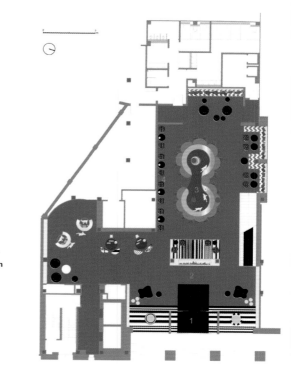

1. Entrance
2. Reception
3. Bar
4. Lounge

1. 入口
2. 接待区
3. 酒吧
4. 休息室

米索尼爱丁堡酒店提供136间客房和套房,在其中一些客房可以欣赏到城市的地标建筑和皇家大道。客房继续沿用黑白色调,整个空间在黑白的底色下不时点缀着其他艳丽的颜色。形式和功能相结合,为了满足现今旅行者的需求,整个住宿环境都精心设计,并搭配精选的米索尼家居系列产品。

每间客房有一间充满意大利风格的双人浴室。浴室包含洗手盆、马桶和配置了长凳、花洒和手持淋浴头的淋浴间。配备的化妆用品由米索尼酒店定制,米索尼品牌香水的味道充满整个空间。

意大利餐厅位于酒店的正中心。在2011年和2012年苏格兰餐厅大奖评选中被评选为年度最佳意大利餐厅。自从开业以来,意大利餐厅屡获大奖,在爱丁堡餐饮业树立了稳固的地位。

融合了时尚和功能性,意大利餐厅是一间充满活力和喧嚣的餐厅,从漂亮的米索尼印花桌布到黑白标志陶器,餐厅装饰全面地展现了米索尼的品牌特色。

受传统的意大利式"休闲"酒吧的灵感启发,米索尼酒吧是爱丁堡必去的目的地。它位于一层酒店大堂。酒吧可以容纳50名客人,酒吧提供的鸡尾酒单中包括经典饮品,无酒精鸡尾酒以及由酒吧定制的酒品。

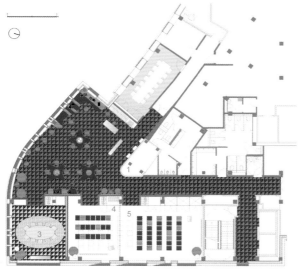

酒店有三间独特充满活力的私人会议活动空间,每一间都经过特别设计,与米索尼酒店的时尚风格保持一致。三个空间面积从28平方米到78平方米不等,其中最大的一间可以分成两个比邻的空间。另一间100平方米的空间可供作为休息区使用,是作为接待和会议间歇的理想场所。

位于酒店三层的米索尼水疗中心提过最新的专业理疗服务,帮助客人舒缓,修复及重振精神。室内装饰着米索尼标志性印花,对传统水疗中心的室内装饰发出挑战,并成功的营造出奢华、舒适的气氛。

| | |
|---|---|
| 1. A corner in reception area | 1. 接待区一角 |
| 2. VIP boardroom | 2. 贵宾会议室 |
| 3. The living room in Lawnmarket Suite | 3. 劳恩市场套房的起居室 |
| 4. The stylish seating area in Rosa Suite | 4. 罗莎套房的时尚起居空间 |
| 5. Bold colours chairs and sofas in breakout area | 5. 休息区的彩色椅子和沙发 |
| 6. A corner in the bar | 6. 酒吧一角 |
| 7. The seating area in Suite d'Argento | 7. 银套房的起居空间 |
| 8. The seating area in Missoni Suite | 8. 米索尼套房的起居空间 |
| 9. The bedroom in Lawnmarket Suite | 9. 劳恩市场套房的卧室 |
| 10. A corner in Piccolo Room | 10. 小客房一角 |
| 11. The bed in Grandioso Room | 11. 大客房的床 |
| 12. The seating area in Grandioso Room | 12. 大客房的起居空间 |
| 13. The bathroom in Missoni Suite | 13. 米索尼套房的浴室 |
| 14. The bathroom in Piccolo Room | 14. 小客房的浴室 |

| | |
|---|---|
| 1. Reception | 1. 接待区 |
| 2. Prefunction area | 2. 会议前厅 |
| 3. Board conference room | 3. 董事会议室 |
| 4. Meeting room 1 | 4. 会议室1 |
| 5. Meeting room 2 | 5. 会议室2 |

# Medusa's Palace

美杜莎的神殿

It's not too exaggerated to describe the Italian famous brand Versace with the word fashion empire. From the original fashion design to jewellery, perfume, cosmetic, to nowadays houseware, and even car and hotel design, Versace exists in each corner of the fashion life. Giani Versace pushed the brand to the top of world fashion with his heart of art and hands of God. The style of Versace he has set up reflects his love to the gorgeous style of Renaissance. Women's dresses with bright colours, which are comfortable and vanward, can also highlight women's wonderful figures and glamour. Just as the brand logo expresses, Medusa who has extraordinary beauty symbolises incomparable beauty and glamour. It is just Versace's design purpose to make the guest feel an irresistible sense of beauty.

This sense of beauty is also reflected in his crossover design. When he was young, Versace had been majored in architecture, which has more or less influences to his clothing design. And there must be some esthetic relationships between his clothing design and architecture design. It is visible to see from his villa design that he has deep love for historic buildings. Just like the clothes he had designed, which is full of literature and charm, his villa was designed to be as gorgeous as the Baroque Age. And the design of Versace Palace Hotel just copies or even surpasses the design concept of his villa. The cup trays, adornments, curtains, beddings, and even coatings and tiles, which are as delicate as the Versace clothes, are all elaborately designed by the designer himself. The whole hotel is full of women's elegance of Versace style, and the Baroque style makes it like a temple, where Medusa's magic actually attacks through the space.

范思哲（Versace）这个意大利的知名品牌如今用时尚帝国一词来形容毫不为过。从最初的时装设计到珠宝、香水、化妆品，再到如今的家居用品，甚至设计汽车、酒店，范思哲如今存在在时尚生活的各个角落。詹尼·范思哲凭借其一颗艺术的心灵以及一双上帝之手将品牌推向世界时尚界的巅峰，他设立的范思哲风格透露出他本人对文艺复兴时期华美风格的热爱，女性服装色彩鲜艳，兼具舒适性和先锋艺术的韵味，又能够凸显女性的美丽身材和魅力。就像品牌的标志所表达出的含义，拥有非凡美貌的美杜莎女神象征着无与伦比的美丽与魅力，正如范思哲的设计目标，让客人感受到难以抗拒的美感。

这种美感也同样体现在他的跨界设计之中。青年时期的范思哲曾经学习建筑专业，这或多或少影响着他对服装的设计，在他的服装设计与建筑设计之间也必然有着审美上的联系。范思哲对古建筑的热爱从他对自己别墅的设计可见一斑，如同他设计的服装充满文艺气息和韵味，他将自己的别墅装饰得充满巴洛克时期的华美。而范思哲宫殿酒店正是对他的别墅设计理念的一次复制，甚至是一次超越。与范思哲服饰同样华美精致的杯盘、装饰品、窗帘、床品，甚至是涂料、瓷砖都经过设计师的精心考量。整体酒店充满了范思哲式的女性优雅，巴洛克的风格让人仿佛来到一座神殿，美杜莎的魔咒在空间之中真实袭来。

# Palazzo Versace

范思哲宫殿酒店

Completion/Renovation date: 2000/2011
Location: Main Beach, Gold Coast, Australia
Size: 200 hotel rooms and suites
Designer: Soheil Abedian, Gianni Versace
Photographer: Palazzo Versace

完成/翻新时间：2000年/2011年
项目地点：澳大利亚，黄金海岸主海滩
规模：200间客房
设计师：索菲·阿巴蒂安（桑德兰集团），詹尼·范思哲
摄影师：范思哲宫殿酒店授权

A waterfront setting, spectacular architecture, Italian furnishings, extraordinary service and an ambience of pure glamour. Palazzo Versace is everything you would expect from the world's first fully fashion-brand hotel; home to three award winning restaurants, famed day spa, Fitness & Wellbeing Centre, private marina, Versace Boutique, meeting and event facilities and access to Australia's first poolside Water Salon cabana experience.

Acclaimed as Australia's premier leisure destination, the Gold Coast is famed for its glorious sub-tropical climate and cosmopolitan enjoyment of life. Nestled between the beautiful Pacific Ocean and the sparkling Gold Coast Broadwater, the hotel inspires an indulgent getaway offering an unmistakably Versace experience.

Palazzo Versace's selection of award-winning restaurants set the standards in cuisine and dinning, from their signature restaurant, Vanitas, providing guests with an extraordinary gastronomic encounter, and Vie Bar + Restaurant showcasing contemporary flavours in a distinctive, stylish space overlooking the Gold Coast Broadwater, to sophisticated casual dining in Il Barocco Restaurant famed for its lavish seafood buffet and classical dishes. In addition to the three acclaimed restaurants, Palazzo Versace also boasts two ultimately chic bars including Le Jardin, where you can try the latest seasons cocktail or relax with a tea or coffee and high tea within a beautiful setting.

Housing 200 light-filled bedrooms and suites in timeless craftsmanship accessorised by Bespoke Versace furnishing, luxurious fabrics, floor-to-ceiling windows, lavish colourful silks and a "Juliet" balcony inviting the Gold Coast sunshine and azure skies to pour in. The similarities between the hotel and traditional palace design can be seen with the opulent foyer and dining areas representing the meeting places of the palace while the accommodation in the left and right wings ("Ala Sinistra" and "Ala Destra") represent the palace residences.

Palazzo Versace is "a destination within a destination", with a glorious sub-tropical climate; cosmopolitan approach to life; and the beauty of a Renaissance palace combined to create the perfect holiday and the ultimate escapism.

滨水景观，宏伟的建筑，意大利式家私，超凡的服务以及极致魅惑的氛围，在范思哲宫殿酒店客人可以在这里找到在世界第一时尚品牌酒店想找到的一切：三间获奖餐厅、著名的日间spa、健身中心、私人码头、范思哲精品店、会议宴会设施以及澳大利亚第一家池边水沙龙。

被誉为澳大利亚首选目的地，黄金海岸以璀璨的亚热带气候以及都市娱乐生活方式著称。位于这样星光熠熠的黄金海岸宽水海滩和美丽的太平洋之间，酒店自然成为奢华逍遥之地，纯正范思哲的体验之地。

范思哲宫殿酒店的精选获奖餐厅成为了烹饪和就餐的典范。招牌餐厅凡尼塔斯餐厅，为客人提供与众不同的美食待遇；生活酒吧餐厅提供当代美食，独特时尚的空间遥看黄金海岸宽水滩；巴洛克餐厅有着精致华丽的休息就餐环境，它以丰富的海鲜自助餐及经典菜式著名；除了这三家盛名的餐厅，范思哲宫殿酒店同样设立了两家极致时尚的酒吧，其中的花园酒吧，在这里客人可以尝试最新的鸡尾酒或者享用一杯咖啡，或者在美丽的布景下细斟一杯下午茶。

在酒店200间精心布置的客房与套房内，室内设计历久弥新，其中布置有定制的范思哲家私、奢华的织挂品、落地窗、色彩丰富艳丽的丝绸以及可以享受黄金海岸阳光及湛蓝天空的"茱莉叶式"豪华阳台。在酒店门厅和大型宴会、会议空间设计上，可以看出范思哲宫殿酒店和传统宫殿设计方面的相似性。同样的，在酒店的左右厢客房同样可以看到宫殿般的居住环境。

范思哲宫殿酒店可谓是旅游景区中的景区，它独享阳光充沛的亚热带气候，大都市生活方式以及文艺复兴时期的宫殿建筑设计之美，毫无疑问是一处完美的度假休闲胜地。

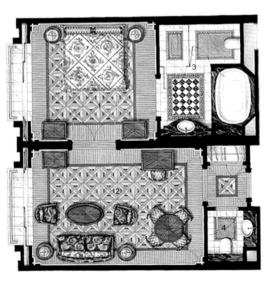

1. The lobby
2. The overview of hotel landscape
3. A detail of exterior
4. A view from the balcony
5. The overview of lobby
6. The entrance to lobby
7. The lounge in the lobby
8. The ballroom
9. The living room in Imperial Suite
10. The bedroom in Superior Room

1. 大堂
2. 酒店景观总览
3. 外立面细部
4. 阳台风景
5. 大堂全景
6. 大堂入口
7. 酒店大堂的休息区
8. 宴会厅
9. 皇家套房客厅
10. 高级客房卧室

| | |
|---|---|
| 1. Bedroom | 1. 卧室 |
| 2. Living room | 2. 客厅 |
| 3. Bathroom | 3. 浴室 |
| 4. Toilet | 4. 卫生间 |

# Original Fashion

独创时尚

Camper, which is a shoemaking brand stemmed from Spain, was born in 1975. The brand's history can trace back to 1877. The traditional Spanish shoemaker Antonio Fluxa had studied industry shoemaking in the UK. And then he led a batch of traditional Spanish shoemakers to set up a shoe factory which combined the traditional shoemaking crafts with the industry mechanisation. It was honoured as a creative and daring challenge at that time. Now Camper is in the control of Lorenzo Fluxa, the grandson of Fluxa, who set up the brand in 1975. During the development of these years, the brand has been keeping the original innovative and distinctive brand concept all the time. From the brand's operation philosophy to its design style, Campter has been moving ahead in the road of its own design without pursuit of fashion and trend. Lorenzo Fluxa had even said, "I will feel uncomfortable if people say that our brand is a fashion brand." In the road of development, Camper never lower its head to the trend, and its design is fresh and natural with rich colours, and personalised pattern designs on the vamp. The patterns full of imagination make Camper shoes fresh and humorous, which also make the guests who are familiar with this brand recognise it at the first sight. The most unique place is Camper's sole. Different poems and fables are printed on the soles of each pair of shoes. Moreover, their comfortability and affinity with people make it the most popular brand. The design prototype of Camper's first type of shoes is the shoes wear by the farmers on a Spanish island. The designs from that on are all based on life. It pays attention to the common people's life and makes the shoes suitable and comfortable. Guests are also ready to start a new day wearing Camper shoes. Since 2003, Camper had brought their innovativeness and affinity with people to their brand hotel design. Two Casa Camper hotels have been opened respectively in Barcelona and in Berlin by now. In Casa Camper Barcelona, 10 bicycles are innovatively hung by the designer on the ceiling of the lobby, which can be associated with the material quality of Camper's sole. The walls are painted with bold red and white just as the Camper shoes, which forms a distinct contract. The guestroom and the living room are innovatively separated with the guestroom on the red wall side and the living room on the white wall side. The hotel also exceeds other hotels in its affinity with people. What is mentionable is the water recycling system that is specially designed for the hotel. It uses the most natural manner to purify the water and prevent it from the pollution of chemicals.

看步（Camper）是源自西班牙的制鞋品牌，诞生于1975年。而说到品牌的历史要追溯到1877年，西班牙的传统制鞋匠安东尼奥·弗拉萨（Antonio Fluxa）在英国学习了工业制鞋，便带领西班牙的一批传统鞋匠们建立了一家将传统制鞋工艺与工业机械化相结合的鞋厂，这在当时被誉为一次创新大胆的挑战。如今的看步由弗拉萨的孙子洛伦佐·弗拉萨（Lorenzo Fluxa）执掌，他在1975年创立了品牌。经历这些年的发展品牌始终保持着最初勇于创新和与众不同的品牌理念，从品牌的经营理念到设计风格，看步都在自己的设计道路上前行，不追逐时尚潮流。洛伦佐·弗拉萨甚至说过："当有人说我们是时尚品牌时，就会使我不舒服。"在这条不向潮流低头的发展道路上，看步的设计清新自然，色彩丰富，他们崇尚鞋面的个性图案设计。富有想象力的图案令看步鞋看起来清新幽默，让熟悉品牌的顾客一眼就能认出是看步鞋。最特别的是看步的鞋底，他们在每双鞋的鞋底印上不同的诗句和寓言。另外，顾客喜欢看步鞋的另一个原因是它的舒适性和亲民性，看步第一款鞋的设计原型是西班牙小岛上农夫穿的鞋，此后的设计也都在生活中取材，它注重普通人的生活，把鞋做得合脚舒适，顾客们也愿意穿着看步鞋展开一天的日常生活。从2003年起，看步同样将这种创新性与亲民性带入到他们品牌酒店的设计中，目前在巴塞罗那和柏林已经开了两家营地之家酒店。在巴塞罗那营地之家酒店，设计师创新地在大堂内的天花板上吊起10辆自行车，让人联想到看步的鞋底材质。墙面同样和看步鞋一样涂上大胆的红色和白色，形成鲜明的对比。酒店还创新的将卧室和客厅分开，卧室在红色的墙一边，而客厅则在白色墙一侧。这家酒店还在亲民性上胜过其他酒店，值得一提的是酒店专门设计的水循环系统，用最自然的方式让水经过净化而免于受到化学物质的污染。

# Casa Camper Barcelona

巴塞罗那营地之家酒店

**Completiong date:** 2003
**Location:** Barcelona, Spain
**Designer:** Fernando Amat from Vinçon, Jordi Tió
**Size:** 20 guestrooms and 5 suites
**Photographer:** Casa Camper Barcelona

完成时间：2003年
项目地点：西班牙，巴塞罗那
设计师：费尔南多·阿迈特，约尔迪·狄奥
规模：20间客房和5间套房
摄影师：由巴塞罗那营地之家酒店提供

Casa Camper is a genuine 100% Camper product and not a licensed brand extension. Twenty-two years after opening the first shoe shop in Barcelona, Camper opened their first hotel in the Raval district in collaboration with Vinçon's Fernando Amat and Jordi Tió.

Casa Camper hotel is an outstanding building strategically located in the cross section between the Ramblas and the MACBA museum in Barcelona. Raval's district is magical, multicultural and creates a unique vibe to the city itself. The area is full of galleries, bars, little restaurants, and only two minutes walking distance from Plaza Cataluña.

The building, a typical 19th-century Gothic tenement, was restored by architect Jordi Tió and perfected by an interior from Fernando Amat of Vinçon. It now stands as one of the city's premier boutique hotels. The fusion between tradition and the avant-garde creates a flavuor in sync with the Camper Values.

Camper created this new unique hotel with the intention of providing its guests with an oasis of tranquillity in the middle of a district quite the contrary, Raval is crowded and bustling with city life.

If one didn't notice the aged illuminated H-O-T-E-L sign, and simply passed by the gallery windows, or walked into the hall, one might think they were in a bicycle store or art gallery.

Looking up, there are 10 bicycles hanging from the ceiling waiting to be used by visiting guests. All intentionally provoking confusion to the passer by, looking in on to this unique space, from the busy street.

Once inside the hotel, the guests are greeted with a photo mural by Hannah Collins. These images show shop fronts from the surrounding neighbourhood.

The central island in the lobby is a clear intention of creating an eclectic mix of function and design, where a luxurious marble and wood reception counter decorated with pop-art contrasts and an antique window display cabinet which houses the guest shop and is filled with alcoholic beverages, bath goods, books and other curiosities available to buy. The counter moves on to another technological space where the building installations are controlled. In the middle of the island you will find a table, similar to one you would find at home, mixing books, flowers and apparently useless items, yet so important to create the atmosphere of form and function.

We are about to enter a world where the graphics have been created by designer América Sánchez and Albert Planas and the messages from a brand that has always promoted walking over running…

Casa Camper Barcelona offers two meeting rooms, "CAM" and "PER" meeting rooms are fully equipped and can adjust to any meeting type of set up.

Depending on the guest's needs, meeting rooms can be joined together to form one space with capacity for 70 people and direct access from Elisabet Street. The terrace on the sixth floor, with its wonderful view of Barcelona, is the ideal location to organise cocktails or informal and creative meetings. A meeting space located on the main floor and furnished with benches, tables and chairs all with the simple idea of functionality, simply a space to have breakfast, a snack or beverage…

Once arriving to your floor, you will find that the walls are painted differently and one side is red, and the other white… On each red door you will find the word "dormitorio" or bedroom, and on the white door, "salita de estar" or living room, each with the same number, as Casa Camper offers each guest two unique spaces, one for sleeping and one for working, resting or just simply hanging out… Each room is separated by the hallway, which separates the two different areas.

The bedroom's are always located on the "quiet side", and look over an extraordinary vertical garden which is composed of 117 aspidistra plants. The plants and their long sinuous shiny leaves are perfectly arranged on an 18-metre-high shelf, and create a relaxing atmosphere. The bathrooms are precisely the area that opens up to this surprising urban vertical garden. If you appreciated the over sized elevator buttons, you will also appreciate the easy to use bathroom knobs, and the privilege to shower, shave and put on your makeup in the natural light of the vertical garden. All bathrooms look out to the garden and provide guests with the natural sensation of vegetation and natural light.

In addition, the high tech water recycling system lets you shower and bath without worries, and rest assured, you know that the designers care, and every time you use the water it goes through a natural

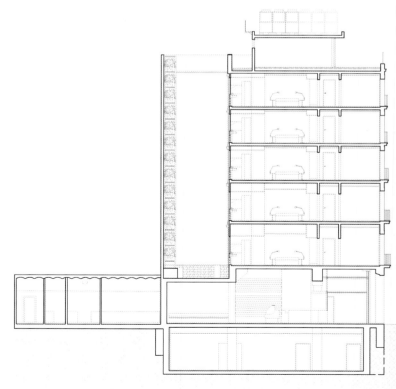

recycling system which allows the water to be purified without chemical treatments. A solar energy water heating system is in use throughout the hotel.

Checking out the "curiosities" that make your space more comfortable and unique: Hygiene is clearly important throughout the hotel therefore there is no carpeting within the hotel. The Shakers designed over 100 years ago the best hanging system ever created. The designers have interpreted their design and have installed this hanging devise throughout the room and they provide guests with a clutter-free room yet at the same time provide them with: A portable lamp for guests' a foldable step ladder to reach higher shelves and hide the luggage, and clear "over forty" size room instructions.

营地之家酒店是100%的营地之家品牌的原创酒店。在看步（Camper）开了他们在巴塞罗那第一家鞋店的22年后，由瓦松设计公司的费尔南多·阿迈特和约尔迪·狄奥合作设计的第一家营地之家酒店在拉瓦尔区开业了。

营地之家酒店坐落在介于兰布拉大道和巴塞罗那当代美术馆之间的十字路口。拉瓦尔区对于这座城市本身是一个充满魔幻色彩、多文化融合，具有独特氛围的区域。这里遍布着美术馆、酒吧、小餐馆，离加泰罗尼亚广场步行只需两分钟。

酒店建筑师一座典型的19世纪哥特式住宅，经由建筑师约尔迪·狄奥的重新翻修和室内设计师费尔南多·阿迈特对室内的完善。如今它成为了这座城市中主要的精品酒店之一。酒店将传统与前卫艺术融合，创作出一种与看步品牌价值相一致的风格。

在这片熙熙攘攘的地区，看步创立的这家新颖独特的酒店目的是能提供给客人一片宁静的绿洲。

如果客人没有注意到酒店那块写着H-O-T-E-L的霓虹标志牌，就匆匆路过橱窗，或者进入大厅，他一定会认为来到了一家自行车店或是美术馆。

向上看，有10辆自行车被挂在天花板上等待游客的使用。这一切都是故意引起那些在繁忙的街道，远望这个独特空间的过路人的错觉。

一旦进入酒店，客人会看到由摄影师汉娜科林斯拍摄的一组壁画。图片的内容展示了酒店周围的店面景象。

设计酒店大堂中心区域的主要目的是营造出一种将功能和设计融合在一起的折衷氛围，在这里奢华的大理石和木制的接待台装饰着流行艺术符号，古旧的橱窗展示着柜仿佛把商店都藏在了店内，里面装满了酒类饮品、洗浴用品、书籍和其他珍品古玩。从柜台将目光移到另一片聚集了装饰品的区域。在这片区域的中心，客人可以发现一个桌子，这与我们在家中的桌子非常相似，层叠的书籍、鲜花和无用的物件，但是却营造出形式和功能兼具的氛围。

接下来进入的区域是一个布满了由设计师阿美莉卡·桑切斯和阿尔伯特·帕拉纳斯创作的插画，图片创意来自于一个倡导步行胜过跑步的品牌。

巴塞罗那的营地之家酒店提供两个会议室，"CAM会议室"（57平方米）和"PER会议室"（5平方米），会议室都配备了完善的设施，适用于各类型的会议。根据客人的需求，两个会议室可以合成一个112平方米的空间，可以容纳70人参会，出口直接面向伊丽莎白街。在六层的天台，壮美的巴塞罗那景色是理想的鸡尾酒会和非正式会议的理想选址。会议空间位于主楼层，这里配备了椅、桌子，椅子兼具简单的功能性，在这里还可以享用早餐、餐点和饮品。

当客人步入客房楼层，会发现这里的墙壁被涂上了不同的颜色，一面是红色，而另一面是色……在每一扇红门上都写着"dormitorio"或者"bedroom"（卧室），在另一边白色的门上着"salita de estar"或者"living room"（客厅）的字样，每一边都标着相同的数字，这是因为地之家酒店为每一位客人提供两个独特的空间，一间供人睡觉，一间供人工作、休息或者仅是简单的聚会……每一间房间由走廊间隔，将两个功能区分开。

卧室都位于安静的一侧，可以看见由117棵叶兰组成的非常特别的垂直花园。这些植物有着长状曲的闪亮的叶子，被布置在18米高的架子上，营造出一种轻松的氛围。浴室是观赏这个令人惊喜都市垂直花园最恰当的地方。如果你感激电梯的大号按钮，那么同样也会惊喜与浴室的大号把手。在这里淋浴、剃须或是化妆都能享受到垂直花园的自然阳光。所以的浴室都朝向花园，并提供给人自然的感受。

另外，高科技的水循环系统让客人放心淋浴、洗浴。水流在客人洗浴时经过自然的循环系统，让不受其他化学的污染得到净化。此外，整个酒店都使用太阳能水加热系统。

房间里有舒适特别的设施，卫生对于酒店来说是非常重要的，因此在整个酒店看不到地毯。吸取100年前震颤派教徒的吊床设计，营地之家酒店将他们的设计改进，安置在整个酒店的客房，此外，酒店为客人提供的是一个自由布置的空间，但是同时也提供以下设施：便携式的手提灯、可叠的梯子、能够到更高的架子以便放置行李，以及房间使用说明。

酒店还为每位客人设置了个性化的客厅，位于大堂的另一边，每间客厅都有自己阳台可以遥望熙的兰布拉大道，有全天候的奇特景象。客厅都配备了等离子电视、互动式电视、蓝牙和WI-FI网接入等。

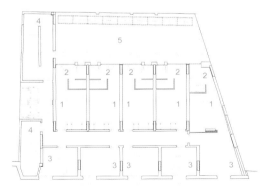

1. Dining area on Terrace
2. The hall featuring the antique window display cabinet
3. The dining area in Tentempie with photos by Hannah Collins
4. The detail for the hanging bicycles in the hall
5. The entrance of hotel
6. The detail of dining area on the terrace
7. Vertical garden
8. The terrace on sixth floor
9. The fitness centre
10-12. The details in fitness centre
13. Meetings and events room
14. The corridor separating the red bedroom and white living room
15. The bathroom
16. A corner in bedroom
17. The personal living room in the hall
18. One of white living rooms
19. The living room with hammock

1. 露台就餐区
2. 大堂内的带橱窗的古典货柜
3. 谭谭皮埃餐厅内挂着有摄影师汉娜·科林斯拍摄的照片
4. 大堂内悬挂着的自行车细节
5. 酒店入口
6. 露台就餐区细节
7. 垂直花园
8. 位于酒店6楼的露台
9. 健身中心
10-12. 健身中心细节
13. 会议活动空间
14. 将红色卧室和白色客厅分开的走廊
15. 浴室
16. 卧室一角
17. 大堂内的个性客厅
18. 其中一间白色客厅
19. 带吊床的客厅

**1. Bedroom**  **4. Staircase**  **1. 卧室**  **4. 楼梯间**
**2. Bathroom**  **5. Terrace**  **2. 浴室**  **5. 露台**
**3. Living room**  **3. 客厅**

# Fairy Shoes Home

仙履家园

Almost everybody has heard the story of "Cinderella", but who knows what Cinderella's dancing shoes look like? People may be suddenly enlightened if they are Ferragamo's shoes. It's never too exaggerated to describe Ferragamo with "Magic Shoes".

Salvatore Ferragamo, who is the founder of this brand, was born in a southern Italy Napoli town Pueblo Bonito. In 1914, Salvatore Ferragamo immigrated to America. And in 1927, he came back to Italy and set up his first specialty store named after his name in Florence. He had won the name of "A Shoemaker Employed by Stars" for his special attention to quality and details. Nowadays, Salvatore Ferragamo is one of the world top designers for shoes, leather products, accessories, clothing and perfume. Salvatore Ferragamo's design style is luxury and elegant with equal stress on practicability and style. He has good reputation in the luxury field from all over the world with his traditional manual design and style.

The inspiration of the design of Salvatore Lungarno Hotel is from Italian Classicism. It is closely linked to the creativity of Ferragamo, which makes the whole tone and lines simple and elegant. The hotel is decorated with many artists' authentic works, classical furniture, and in the meantime equipped with modern practical high-tech facilities. Each piece of furniture which shows a kind of luxury and elegance in the luxury suite is produced by Ferragamo. The classic style of columns, walls, and the classic lines which divide the space clearly draw the outline of the essence of Classicism. Just like the shoes and clothes designed by Ferragamo, no matter how boundless his creative designs are, they all give a feeling of luxury and elegance without exception. This may be the Neoclassicism in the fashion field. Both novel sculpts and fresh materials are the necessity to adapt to the development of the times. However, Ferragamo can refine those classic design elements and perfectly infuse them into the new design. That's why it can be accepted. Luxury and elegance are the common features of both Lungarno Hotel and Ferragamo.

Where will the stories happen when the regular customers such as Audrey Hepburn and Sophia Rossi gracefully put on their precious "crystal shoes"? It seems that the only answer is Lungarno.

"仙履奇缘"的故事人人都听过,而谁知道灰姑娘的舞鞋到底是什么样子的呢?如果说是菲拉格慕的鞋品,人们也许会恍然大悟,"仙履"用来形容菲拉格慕再合适不过了。

品牌的创始人萨瓦托·菲拉格慕,生于意大利南部拿波里小镇波尼托。1914年,菲拉格慕移民到美国。1927年,萨瓦托·菲拉格慕回到意大利,在佛罗伦萨开设以其同名品牌的首间专门店。萨瓦托·菲拉格慕异常关注质量和细节,因此他赢得了"明星御用皮鞋匠"的称号。而今,萨瓦托·菲拉格慕是皮鞋、皮革制品、配件、服装和香水的世界顶级的设计者之一。菲拉格慕(Salvatore Ferragamo)设计风格:高雅华美,兼顾实用性和款式。菲拉格慕坚持传统手工艺和款式,品质享誉世界时尚界。

菲拉格慕朗伽诺酒店设计的灵感来自于意大利新古典主义,并与菲拉格慕的创造力紧密相连,整体色调和线条简练优雅。酒店装饰众多艺术家真迹,古典家私同时配备实用的现代高科技设备。豪华套房内的每件家具都是菲拉格慕出品,流露出一种华贵典雅的风格。古典式样的立柱、墙围、空间分割直至经典的线条,虽不繁复但却清晰的勾勒出古典主义的本质。就像菲拉格慕的鞋品、服饰,无论设计创意如何无穷,却无一例外都给人华贵高雅的感觉。这或许就是时尚领域的新古典主义。新颖的造型,新鲜的材料,这都是适应时代发展的必然,但菲拉格慕能够提炼出那些经典的设计元素完美的将他们融入到新的设计之中,这大概是它取得认可的原因之一。华贵、优雅是朗伽诺酒店和菲拉格慕鞋品共同的特点。

奥黛丽·赫本、索菲亚·罗兰是菲拉格慕的拥护者,当她们优雅高贵的穿上珍爱的"水晶鞋",那故事发生的地点应该是哪里呢,好像也只有朗伽诺这一个答案。

**Completion/Latest renovation date:** 1997/2012
**Location:** Florence, Italy
**Size:** 43 guest rooms
**Designer:** Michele Bönan, Nino Solazzi
**Photographer:** Lungarno Collection

完成／翻新时间：1997年/2012年
项目地点：意大利，佛罗伦萨
规模：73间客房
设计师：米歇尔·博南，尼诺·索拉奇
摄影师：朗伽诺酒店集团

# Hotel Lungarno

朗伽诺酒店

Owned by the Ferragamo family and the flagship property of the Lungarno Collection, Hotel Lungarno is a restoration of a 16th century grand residence and is located in one of the most extraordinary locations of Florence, with views that overlook the Arno River and the iconic Ponte Vecchio. Opened following a refurbishment in 1997, the hotel stands out as one of the most desirable destinations in the city for international clientele, with a magical atmosphere of timeless hospitality and inviting charm. Interior design by architect Michele Bönan is inspired by the nuova classicità, with a smart balance of ivorytoned fabrics and ocean blue carpets with elegant details and antique furnishings that create the ambience of an aristocratic Italian residence.

Arrivals are welcomed into a warm lobby. Over 400 works of art including those by Cocteau, Picasso, Bueno and Rosai adorn the walls of public areas and each of 73 guest rooms and suites. A design by Picasso set above the central fireplace can also be found in guest room living areas. Guests enjoy privileged access to Borgo San Jacopo Ristorante – the city's "dining room on the river" – with seasonal menus that draw upon the tradition of the great Italian kitchen. Onsite meeting space includes Sala Guarnieri, while just off-site the hotel operates Palazzo Capponi with the prestigious Capponi Suite apartment and the Sala Poccetti reception hall with 16thcentury frescoes and adjacent annex rooms for a range of events.

### GUEST ROOMS

Hotel Lungarno has 73 guest rooms including 13 suites. Pampering and luxurious, all accommodations look out through large windows and individual terraces onto views of the Arno River and ancient Florence. Interior design is inspired by the nuova classicità, with ivory-toned fabrics, ocean blue carpets, antique furnishings and 20th century works of art primarily from Italian artists. From silver trays arranged with Italian fragrances, small but elegant details add touches of luxury, while in-room technologies such as fax and sensory-activated controls for lights, air-conditioning and heating assure convenience. Spacious marble bathrooms have large shower cabins with nickel taps and fittings.

### DELUXE JUNIOR SUITES

The heights of accommodation at Hotel Lungarno, each of four Deluxe Junior Suites has a character all its own. Spacious living areas and bedrooms are laid out with attributes including original paintings, antique furnishings, double bed, sofa bed, the latest technological conveniences, and floor-to-ceiling sliding doors that open onto balconies or terraces with river and city views. Each is part of the Lungarno Suites collection and all can be joined with additional rooms to create large suites capable of accommodating up to six people.

### DINING

Borgo San Jacopo Ristorante offers an exclusive riverside setting for savoring extraordinary traditional Tuscan home cooking and regional specialties, as well as over 500 labels of regional and international wines. Opened in 2004 and named one of the best restaurants in Florence by L'Espresso restaurant guide in 2011, the 65-seat restaurant occupies two levels overlooking the Arno River with rare Ferragamo photos that dot the walls.

3

1. Suite Picasso Interior
2. Lounge Bar Picteau
3. Suite Torre interior
4. Hotel Lungarno Lobby
5. The layout of Lounge bar
6. Deluxe guestroom interiors with view on the Arno River
7. Hotel Lungarno Lobby with its own Picasso
8. Ponte Vecchio view from the lobby
9. The entrance to Restaurant Borgo San Jacopo

1. 毕加索套房内饰
2. 比克图酒廊
3. 托雷套房内饰
4. 朗伽诺酒店大堂
5. 酒廊布局
6. 能欣赏到阿诺河景色的豪华客房内饰
7. 酒店大堂内装饰着毕加索画作
8. 大堂的维奇奥桥景
9. 圣雅各布餐厅入口

1. Meeting room
2. Gallery
3. Reception
4. TV room
5. Toilet

1. 会议室
2. 画廊
3. 接待处
4. 录像室
5. 洗手间

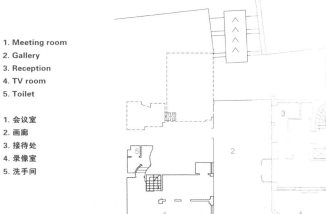

Architect Michele Bönan took inspiration from 1950's glamour, the Italia "new deal" and the "Made in Italy" movement so closely tied to the creativit of Salvatore Ferragamo. Walls are embellished with reproduced designs sketches and images of this fashion epoch. An arched window on eac level provides elongated architectural perspectives on the river belov Tables are set with crystal glassware and fine porcelain. Custom lightin reflects the atmosphere of a 1950's film set, arranged to highlight the tabl and décor without obscuring views of other diners. There are two gran mirrors positioned opposite one another in the front of the mezzanine are creating the appearance of infinite space. There are also soft linen and cotto furnishings in white tones with grey mineral notes that play off oak floors an black and brown details.

## BAR

Picteau Lounge Bar is favoured by convivial international clientele. Laid ou like a living room overlooking the Arno River, the lounge fills with sunlight an moonlight through great windows, giving the sensation of sailing gently ove the water. Guests relax in antique furnishings before a fireplace and origina paintings by Picasso, Rosai and Cocteau.

## MEETINGS

Located near the entrance to the hotel, Sala Guarnieri is an inviting space fc private meetings, dinners and events. The Sala Guarnieri has a capacity fc 30 people before large, light-filled windows with views of the River Arno an Ponte Vecchio. Highquality, contemporary presentation technologies and on site catering and professional support staff assure success for a wide rang of functions. Private parties can also be arranged onsite at Borgo San Jacop Ristorante and Picteau Lounge Bar.

作为菲拉格慕家族的产业,同样是朗伽诺酒店集团旗舰酒店的朗伽诺酒店是由一座16世纪的奢华民居翻修而成。酒店坐落在佛罗伦萨最非凡的地区,在这里可以遥望阿诺河和标志性的维奇奥桥。1997年酒店经过翻修正是开幕,它永恒的热情和诱人的魅力仿佛让整个空间充满魔力,被国际游客认为是佛罗伦萨城中最渴望光临的目的地之一。由建筑师米歇尔·博南为酒店进行室内设计,他的灵感来自于意大利新古典主义,象牙色的织物与海蓝色的地毯组成一种巧妙的平衡感,外加优雅的细节设计与复古家具一起营造出贵族气派的意大利式家居风格。

刚到达酒店,温暖的酒店大堂就迎接着客人。超过400件艺术品装饰着公共区域和73间客房及套房的墙面,包括考克多、毕加索、布埃诺和罗萨伊的作品。其中,毕加索的作品可见于中央壁炉之上,另外在客房的客厅中也会欣赏到这位大师的作品。客人们享有直通本城市"河上就餐室"之称的布尔戈圣雅各布餐厅(Borgo San Jacopo Ristorante)的特权,这里的应季食品菜单集萃了优秀的意大利传统烹饪菜式。酒店的室内会议空间包括古埃尼埃里会议室(Sala Guarnieri),在酒店外可提供的宴会场所包括卡波尼宫殿(Palazzo Capponi)和声名显赫的卡波尼套房公寓(Capponi Suite),此外装饰着16世纪壁画的波切提会议室(Sala Poccetti)和邻近附加的房间能满足一系列宴会活动的需求。

## 客房

朗伽诺酒店有73间客房,包括13间套房。内部富丽堂皇,配有超大的窗户和独立阳台,客人在这里可以欣赏到阿诺河以及古老的佛罗伦萨城的景色。客房的室内设计灵感来自新古典装饰主义,装饰着象牙白织物、海蓝色地毯、古董家具以及来自20世纪最主要的几位意大利艺术家的画作。从散发着意大利香气的银色托盘,小巧优雅的装饰细节都强调了奢华的韵味。此外,房内的设备,例如传真机、感应灯、空调和供热装备,确保了客人的方便。宽敞的浴室配有宽敞的淋浴间、镍制水龙头和洗浴配件。

## 豪华套房

在朗伽诺酒店,每一间豪华套房都有着它们自己的特色。宽敞的客厅和浴室按功能精心布置了家具和装饰品,包括原创画作、古典家具、双人床、沙发床、高科技的便利设施,在可以远眺阿诺河和城市景色的露台上安装的滑动门。这其中的每一件都是朗伽诺套房家私系列的产品,所有的家具组合在一起可以满足最多6个人居住的大套房的需求。

## 就餐环境

布尔戈圣雅各布餐厅(Borgo San Jacopo Ristorante)提供了一种特有的河边就餐环境,在这里可以品尝到非同寻常的托斯卡纳家常菜和地区特色菜,同样的还有500种当地和国际品牌的名酒。在2011年出版的餐厅指南杂志L'Espresso中,这家2004年营业的餐厅被评为佛罗伦萨最好的餐厅。餐厅占据了酒店两个楼层,共设65个座位,在这里用餐可以欣赏到阿诺河的美景,同时墙上稀有的菲拉格慕照片印证着这家酒店的时尚背景。

建筑师米歇尔·博南从20世纪50年代光辉的意大利"新政"运动以及"意大利制造"运动中获得灵感,与菲拉格慕的创造力紧密相连。墙面用充满当代时尚感的设计作品、素描和图片装饰。每一层楼都有一个拱窗提供了延长的建筑视角直至阿诺河面。桌子上摆放的是水晶玻璃器皿和精致的陶瓷制品。定制的照明设施使整个空间极像50年代电影场景的感觉,同时可以起到强调同桌就餐者和桌面物品的作用,这样可以避免模糊其他就餐者的视线。在包厢区域的前方,两面大镜子背对背被安放在这里,营造出无限的空间感。柔软的亚麻布及棉质家私采用白色稍加灰色的色调突出橡木地板的黑棕色细节。

## 酒吧

伦卡诺酒吧休息室(Lounge Bar Lungarno)受到喜爱交际的国际客人的喜爱。酒吧的空间被布置成客厅的样式,在这里可以看见阿诺河,同时休息室的大窗让日光和月光在不同的时间填满整个空间,给人一种在海上逍遥航行的错觉。客人们在摆放着古典家私、火炉以及毕加索、罗萨伊、考克多真品的空间内休息。

## 会议空间

在酒店入口的附近,古埃尼埃里会议室(Sala Guarnieri)就在这里,这是一个适合私人会议、宴会,和活动的魅力之地。会议室内可以容纳30人,空间宽敞,灯光明亮,透过窗户还可以欣赏到阿诺河和维奇奥桥的风景。高质量、现代的展示技术与现场贴身专业的服务的团队确保开展各项活动的成功。另外,在布尔戈圣雅各布餐厅(Borgo San Jacopo Ristorante)和伦卡诺酒吧(Lounge Bar Lungarno)还可以举办私人派对。

# The Portrait of Ferragamo

菲拉格慕的自画像

During the war, Ferragamo used Rufiyaa leaves fibre, cork wood and other simple and crude materials to realise his originality. The invisible sandals which is an empirical design pushed out in 1947 helped Ferragamo win the prize of "Neiman Marcus" — the Oscar in fashion world. In 1966, Ferragamo designed a pair of velvet ankle boots for the female star Brigitte Bardot. It can be called a stroke of genius. Different materials are chosen to make Gancino bag in different seasons. In 1990, a bag made of plexiglass was really an eye-opener.

The theme of the Portrait Suites is to commemorate Roman film, art history and Ferragamo's life. The hotel is decorated with photos of many film stars and celebrities in the fashion world. The interior design skill is chic and daring while the furniture is modern and fashionable. The designer decorates the hotel by using a non-traditional skill, for example, the chic table is matched with pigskin tablecloth and assorted lamps. It's the same to Lungarno Hotel that Ferragamo's exquisite is shown through the strong contrast and collision of different colours. And the difference to Lungarno Hotel is that the Portrait Suites reflects Ferragamo's another aspect in design — daring to break the routine and specifications farsightedly. It may be the fine tradition of Ferragamo's brand.

Walking around this autobiographical hotel, you'll find that the photographs for film stars, fashion stars and city sceneries are hung in each corner. The whole building is like an informal family album and tells the legendary story of Ferragamo's fashion family. In 14 different rooms, this kind of fashion character is continuously passed on.

在战争期间，菲拉格慕用拉菲亚树叶纤维和软木等简陋材质演绎其创意；1947年推出的经验设计——"隐形"凉鞋为菲拉格慕赢得了时尚界的奥斯卡奖——"Neiman Marcus"奖；1966年菲拉格慕为女明星碧姬·巴铎 (Brigitte Bardot) 设计了一双天鹅绒及踝短靴堪称神来之笔。菲拉格慕的经典Gancino包每季都选用不同材质制作，1990年的一款更是采用树脂玻璃制造，着实让世人大开眼界。这可谓是菲拉格慕品牌的优良传统。

肖像套房酒店是菲拉格慕旗下的又一家酒店，它的设计主题是纪念罗马电影、艺术史和回顾菲拉格慕的设计生平。在设计上仍然尊重品牌的传统，酒店装饰着许多电影明星和时尚界名人的照片，室内设计手法别致大胆，家私时尚现代。设计师使用一种非传统的手法装饰，例如在别致的餐桌上搭配猪皮桌布和配套的灯具。与朗伽诺酒店一样，菲拉格慕的精致通过不同颜色强烈的反差碰撞凸现出来。而与朗伽诺酒店不同的是，肖像套房酒店体现的是菲拉格慕设计中敢于打破常规，锐意突破定式的一面。

徜徉在这座自传性质的酒店，每个转角都悬挂有电影明星、时尚明星以及城市风光的照片，整个建筑就像一部非正式的家族相册，讲述着菲拉格慕时尚家族的传奇，在14个迥然不同的房间里，这种家族的时尚品格被不断地延伸传递。

# Portrait Suites

肖像套房酒店

Designed to bring more dolce to the lives of guests, the Portrait Suites commemorate Rome's cinema, artistic past and life of Salvatore Ferragamo. Located throughout the upper floors of the townhouse directly above the Salvatore Ferragamo Men's Store on Via Condotti, the hotel offers a unique style of bespoke hospitality, intimacy and privacy, with just 14 luxury suites and studios.

Guests arrive through an elegant and discreet wooden door and the reception area is located on the mezzanine floor. Photographs of film and fashion celebrities line the walls and views of the city and striking urban architecture greet guests at every turn. Personalised butler service is available and caters to each guest, providing them with white glove service and offerings such as private shopping excursions or visits to historic Roman sites. The hotel terrace features an impressive open fireplace in a striking windowed wall surrounded by contemporary furnishings. The hotel's 360-degree rooftop lounge is similarly inviting, with wooden furnishings and lush plants, creating a nice spot for cocktails before breathtaking city views of the Spanish Steps, Trinita dei Monte and more. The rooftop is one of the highest points on Via Condotti.

Interior design by Architect Michele Bönan uses marble and wood in unconventional manners such as boar skin coverings on special tables with incorporated lamps; furniture fabrics inspired by the art of tailoring; and curtain linings that could be mistaken for silk scarves. Like Bönan's design, furnishings throughout are sleek and contemporary. As with all Lungarno Hotels, sophistication is accented with dashes of colour, including cream, pale grey, cyclamen, acid green and warmer wood and leather tones.

**GUEST ROOMS**

Portrait Suites has 14 guest accommodations, many marked by expansive living spaces and all with bar-kitchenettes including refrigerator, microwave oven, espresso/cappuccino machine and dishwasher. In-room technologies were selected with the needs of contemporary travelers in mind, with complimentary Wi-Fi, LDC satellite TV, DVD player, CD player with iPod cable connection, and individual climate control standard. Spacious marble bathrooms have large shower cabins with nickel taps and fittings. Several accommodations have private terraces or balconies, while two suites are outfitted with a sauna and fitness corner.

**DINING/DRINKING**

While all accommodations feature bar-kitchenettes stocked with snacks, Portrait Suites also offers guests daily continental breakfast, available in-room or at the rooftop lounge. The hotel terrace features an "honour bar" with a selection of wines and champagnes.

The rooftop lounge features a full-bar with bartender service. Room service is also available.

**Completion/Latest renovation date:** 2006
**Location:** Rome, Italy
**Size:** 14 suites
**Designer:** Michele Bönan
**Photographer:** Lungarno Collection

完成／翻新时间：2006年
项目地点：意大利，罗马
规模：14间套房
设计师：米歇尔·博南
摄影师：朗伽诺酒店集团

肖像套房酒店的建立是为了纪念罗马电影与菲拉格慕的艺术贡献，它将更多的柔美带入客人所处的酒店环境之中。酒店就坐落在康多提大道菲拉格慕男装店楼上的联排别墅内，它提供的是一种独特风格，在酒店的14间奢华套房和工作室中，客人在这里享受到的是好客、亲切且私人的定制服务。

客人们到达酒店，穿过一扇优雅且素雅的木质门，接待区就位于这层阁楼之中。电影明星和时尚界名人的照片挂在墙上，城市的美景以及令人印象深刻的都市建筑在每个转身的一瞬间候着客人。酒店提供白手套服务和其他服务，例如私人购物短程旅行或者安排参观罗马古迹，个性化的管家服务，迎合每一位客人的需要。酒店的露台装饰着一个令人印象深刻的开放壁炉，它嵌在一面惊人的玻璃墙上，周围摆设这现代家具。

室内设计有建筑师米歇尔·博南执掌，采用大理石和木材作为主料，以一种非传统的手法装饰，例如在别致的餐桌上搭配猪皮桌布和配套的灯具；家纺受到缝纫艺术的灵感启发；窗帘的内衬可能会被误认为是丝绸围巾。像博南的设计一样，随处的家私都显得时尚和现代。和其他朗加诺的酒店一样，精致由颜色的碰撞凸显出来，酒店采用的色调包括奶油色、白灰色、仙客来粉、水晶绿以及暖木色和皮革色调。

### 客房

肖像套房酒店有14间客房，大部分都被标记为奢华的居住空间，所以的客房都是配备酒吧厨房间的，里面设施包括冰箱、微波炉、浓缩咖啡/卡布奇诺咖啡机和洗碗机。客房内的高科技设备经过精心的挑选以满足现代旅行者的需要，包括免费Wi-Fi，LDC卫星电视，DVD播放机，搭配iPod 有线连接的CD播放机，以及个人室温调节装置。宽敞的大理石浴室有超大的淋浴间及镍制的水龙头和毛巾架。几间客房还有露台和阳台，其中两件套房还配有桑拿房和健身区域。

### 餐饮

除了这些能存储零食的酒吧式小厨房，肖像套房酒店还提供给客人日间欧式早晨，客人可以在室内或是屋顶休息区就餐。酒店的露台还设有一间"荣誉酒吧"，提供一系列精选葡萄酒和香槟。在屋顶休息室可以享受到酒吧式的酒保服务。客房服务也是提供的。

1. Private dining area on the terrace
2. Trinita Dei Monti view from the terrace
3. The living room in Deluxe Studio
4. The bedroom in Deluxe Studio
5. The details in guestroom
6. Main reception in Portrait Suites
7. The bedroom in Portrait Studio
8. The details in living room
9. The interior in Penthouse Trinita dei Monti view
10. The interiors in the living room of Via Condotti Suite
11. The bathroom interiors in a suite

1. 露台上的私人就餐区
2. 从露台上看三一教堂
3. 豪华工作室套房客厅
4. 豪华工作室套房卧室
5. 客房内的细节
6. 酒店主接待区
7. 肖像工作室套房的卧室
8. 酒店休息区细节
9. 三一教堂阁楼套房内饰
10. 康多提大道的客厅内饰
11. 酒店浴室内饰

# The Eclecticism in Cartagena

芭蕾舞者的摇滚天堂

Silvia Tcherassi, who is an Asian Colombian fashion designer, has become one of the hottest stars in the fashion world in Milan, Paris, Berlin, Barcelona, London and New York. During 15 years development, she has a style of her own. Her works are evaluated by the critics to be full of eclecticism. She can always use proper exaggeration to infuse the traditional innovation with energy. Nowadays, Silvia Tcherassi transfers her talents into the field of hotel design. The first Tcherassi hotel constructed in Columbia, which directly introduces Tcherassi's design style into hotel design, is a successful leap of crossover. Tcherassi Hotel and Spa is located in the Cartagena city which is a world cultural heritage, where the architectures are typical Columbian colonial style. Tcherassi put on this hotel a modern fashionable inside. All the beddings in the guestrooms are designed by herself, which softens the colonial style of the hotel building. This kind of innovation is a typical express of eclecticism. A delicate grace like a ballet dancer as well as an unruly feel like a rock star can be found in her fashion hotel.

西尔维娅·切拉西是哥伦比亚裔的时装设计师，目前已经成为在米兰、巴黎、柏林、巴塞罗那、伦敦和纽约时装界最炙手可热的新星。经历15年的发展，她的设计风格自成一体，评论家评价她的作品充满了折衷主义感觉。她总能用恰到好处的夸张将传统创新，注入新的活力。如今的西尔维娅·切拉西将她的才能施展到酒店设计领域，第一家切拉西酒店在哥伦比亚的建立将切拉西的设计风格直接引入酒店设计之中，成就了一次跨界的超越。切拉西酒店及水疗中心位于被列入世界文化遗产的城市卡塔赫纳，这里的建筑是典型的哥伦比亚殖民地风格，切拉西将这家酒店换上了现代感的时尚内里，客房用到的床品都出自设计师的亲手设计，这使酒店建筑的殖民风格柔和了许多，这种创新正是典型折衷主义的表现，在这里即能找到犹如芭蕾舞者的精致柔美，又能找到摇滚明星般的不羁。

# Tcherassi Hotel + Spa

切拉西酒店及水疗中心

**Completion/Latest renovation date:** 2009
**Location:** Cartagena, Colombia
**Size:** 7 rooms
**Designer:** Silvia Tcherassi
**Photographer:** Antonio Castenada, Pablo Garcia and Ilan Segal

完成／翻新时间：2009年
项目地点：哥伦比亚，卡塔赫纳
规模：7间客房
设计师：西尔维娅·切拉西
摄影师：安东尼奥卡斯塔涅达，保罗加西亚，伊兰西格尔

Tcherassi Hotel + Spa is a seven-room luxury boutique hotel in a restored 250-year-old colonial mansion in the heart of Cartagena, Colombia's Old City. Designed by famed fashion designer Silvia Tcherassi, the hotel boasts a 50-seat Italian-inspired restaurant named VERA, a recently expanded Aqua Bar, three pools, a full spa and a vertical garden with over 3,000 local plants. Fashion is infused throughout the hotel starting with each of the seven rooms being named after various fabrics used in Tcherassi's collections in addition to guests receiving hand-made fabric flower pins as a gift.

The hotel reignites the magical setting of a bygone era while delivering all of the modern amenities and comforts discerning travellers have come to expect today. Although The Tcherassi Hotel + Spa is an oasis within itself, it is the perfect place from which to enjoy the vibrant colours and culture of Cartagena de Indias. All of the rooms have been designed by Silvia Tcherassi and named for elegant fabrics used in her fashion collections; providing just one example of how Tcherassi incorporates fashion into the hotel's design. Offering either views of the city, the lush garden and beyond, the large rooms (ranging in size from 400 sq ft to 1200 sq ft) feature restored original stone walls, dramatic high ceilings and private walk-out balconies creating an open and airy Caribbean feel. With 30 ft ceilings, wood floors, large soundproof windows and a luxurious open bathroom featuring a double sink, rainshower and elegant tub, there is plenty of room for guests to feel at home

Vera, a 40-seat italian restaurant and lounge, whose name means "truthful" reflecting its authenticity, will satiate diners during breakfast, lunch and dinner seven days a week.

The restored, original 250 year old stone wall and a vertical garden consisting of over 3000 local plants provide diners with a perfect framework to sit and enjoy Cartagena's best new restaurant.

This magnificent scenario will offer light fare including snacks and sandwiches throughout the day atop the hotel's roof deck, all served with a side order of unbeatable 360 degree views of the city and the sea.

Rooted in a holistic approach, The Spa has partnered with Spain's Germaine de Capuccini, which has been sharing its beauty secrets and techniques with consumers and spas for nearly 50 years. With more than 30 treatments, the Tcherassi Spa is truly a haven for relaxation and rejuvenation.

切拉西酒店及水疗中心是一家拥有7间客房的奢华精品酒店，它位于哥伦比亚旧城——卡塔赫纳中心，原址是一家有着250年历史的旧殖民时期公馆，酒店由此翻修而成。 由著名的时尚设计师西尔维娅·切拉西设计，酒店包括一家可容纳50人就餐的维拉餐厅，最新扩建的水晶蓝酒吧，三个游泳池，全套水疗中心和一个种植了3000种本地植物的垂直花园。从以切拉西使用过的服装面料命名的7间客房，到客人收到的礼物——手工布艺花胸针，时尚从头至尾蔓延在酒店之中。

酒店让人想起旧时代的魔幻故事中的场景，同时提供全套现代化设施，让来到这里的挑剔游客充满期待，感到舒适。尽管切拉西酒店及水疗中心是一片绿洲，但是对于来到这里的客人来说，这也是感受活力色彩和卡塔赫纳土著文化的完美场所。酒店内所有的房间都由时尚设计师西尔维娅·切拉西亲自设计，使用的优雅织物也都来自设计师同名织物系列，这次设计提供了一个切拉西是如何将时尚融入酒店设计之中的例证。苍翠繁茂的花园为客人们提供了欣赏城市风景的最佳视角，超大的客房（大小从400平

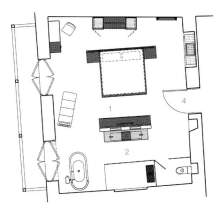

方米到1200平方米）由翻新过的原始石墙围绕，引人注目的高天花板和私人可步入式阳台让客人自由呼吸新鲜空气，营造出开放的加勒比风情。30英寸高的天花板，木质地板，超大隔音窗和奢华的开放式浴室，配置了双人洗手池的浴室，和优雅的淋浴间和浴缸，这样的客房让客人感到宾至如归。

维拉，一间容纳40个座位的意大利餐厅及酒廊，名字的意思为"真诚"，寓意餐厅的可靠性，餐厅为客人提供每周7天早、中、晚三餐服务。

经过翻修有着250年历史的石墙，以及由3000种当地植物组成的垂直花园，提供给就餐者完美的视角，坐在卡塔赫纳最好的新餐厅中欣赏着城市风景是绝佳的享受。在酒店的屋顶，场景装饰奢华并且全天提供便餐，包括零食和三明治，此外还可以欣赏360度的城市风光及海景。

酒店的水疗中心采用整体分析疗法，并与西班牙著名美容品牌纤蔓绮丽合作，与酒店客人分享他们历经50年客户检验的美容秘方和美容技术。切拉西水疗中心提供30多套美容疗法，是名符其实的放松身心、恢复青春的天堂。

| | | |
|---|---|---|
| 1. Bedroom | 1. 卧室 | |
| 2. Bathroom | 2. 浴室 | |
| 3. Balcony | 3. 阳台 | |
| 4. Entrance | 4. 入口 | |

1. Hotel exterior at night
2. Gazar private terrace
3. Roof pool
4. Tcherassi pool in the Garden
5. Spa relaxation area
6. The entrance to the hotel
7. The designer – Silvia Tcherassi in the lounge of the hotel
8. Gazar stairs
9. Mouseline tub
10. Vera Gran Salon
11. The bedroom in Ziberline
12. The reception
13. The bedroom in Organdie
14. Front gate
15. The bedroom with bathroom in Mouseline

1. 酒店外观夜景
2. 嘉泽客房的私人露台
3. 屋顶游泳池
4. 花园内的切拉西泳池
5. 水疗中心的休息区
6. 酒店入口
7. 酒店设计师——西尔维娅·切拉西在酒店大堂休息
8. 嘉泽套房内楼房
9. 莫斯科套房浴缸
10. 维拉·格兰沙龙
11. 姿柏琳客房卧室
12. 接待处
13. 欧兰迪客房卧室
14. 前门
15. 莫斯琳客房的卧室和浴室

# Timeless Classic of Fashion
时尚演绎的永恒经典

All the hotels collected in Part Three present a typical design style admired by fashion designers. Usually, when designing fashionable clothes they absorb the cream of the crop of classic works and then make innovations. This style is true of the hotel design. Boasting a long history and traditional brand effect, Claridge's, Alma Schlosshotel im Grunewald and The Beverly Hills Hotel and Bungalows are reborn via the involvement of fashion and in the meanwhile, fashion finally becomes timeless in the classic decorations of hotels. In Claridge's, a hotel famous for art deco in London, the fashion legend Diane von Furstenberg perfected interior integrating with fashion attitude, fashion concept and art deco; In another magnificent palace architecture, the interior is renovated by the Julius Ceasar in fashion industry – Karl Lagerfeld. The classic charm of the building itself is not affected by the intervention of fashion elements, just like delicate glamour and enduring reputation of Chanel, which is really replenish and timeless; the Beverly Hills Hotel and Bungalows is one of the famous hotels belonging to Dorchester Collection, which is the most stylish hotel management group in hospitality industry. In recent years, it has shined a spotlight because of its renovation and become the inspiring palace for more celebrities, stars and designers in fashion industry.

在这里，酒店体现了时尚设计师推崇的一种设计风格。时尚设计师在设计服装作品时，通常是在经典的作品中吸取灵感并加以创新，这在酒店的设计中仿佛也沿用了这一点。克拉里奇酒店、阿尔玛格吕内瓦尔德皇宫酒店和贝弗利山庄酒店及别墅，这三家具有历史和传统口碑的酒店经由时尚之手孕育出了第二次生命，而时尚也在经典的装潢中得以永生。在以装饰艺术著称的伦敦克拉里奇酒店，时尚传奇黛安·冯·芙丝汀宝用她永恒的时尚态度、时尚标签与装饰艺术完美的融合；另一座华丽如宫殿般的历史建筑，室内由"时界界的凯撒大帝"卡尔拉格菲尔德翻新设计。建筑本身的古典风韵没有因时尚元素的介入而消弱，正犹如香奈儿经久流芳的精致魅力，验证着历久弥新，也验证了风格会永存的定理；贝弗利山庄酒店及别墅是酒店业最具时尚气息的多尔切斯特酒店集团的著名酒店之一，近几年的翻修让它受到更多的瞩目，成为越来越多名人、明星和时尚界人士的灵感之地。

# A Gem of Art Deco Wearing Wrap Dress

穿上裹身裙的装饰艺术瑰宝

Always a showcase for British design talent, Claridge's commissioned world-renowned fashion designer Diane von Furstenberg to create a series of new interiors. The Grand Piano Suite effortlessly blends Claridge's glamour with the elegant style of the legendary Diane von Furstenberg. With marble fireplaces, sumptuous fabrics, hand-selected furniture and signature prints, she has firmly put her fashionable stamp on Claridge's with this ever so stylish suite.

Once we mention Diane von Furstenberg, we'll be immediately associated with her symbolic design of wrap dress which symbolizes elegance and consistence. Claridge's Hotel, which is famous for its Art Deco, is honoured as a gem of London Art Deco. DVF puts her signature wrap dress on this timeless gem, which is still elegant, but more vigour.

As you walk through the grand doors, the Grand Piano Suite makes a stunning first impression – from the spacious entrance hall your eyes are drawn to the magnificent golden hues, a black lacquer console and bespoke leopard skin patterned rug. A black leather-topped private bar opens up like a magic box to reveal a fully stocked selection of classic drinks and delectable treats.

The Grand Piano Suite preserves Claridge's unique heritage whilst showcasing Diane von Furstenberg's influence. The original Claridge's furniture, including a beautiful grand piano that takes centre stage in the living room, and the décor are further complemented by striking lighting in Murano glass.

The master bedroom has a soft colour scheme and state-of-the-art lighting system while the second bedroom, with its own entrance from the hallway, takes its cue from the fashion world with attractive China Club curtains, bright colours and patterns made into bold prints – including an armchair in Diane von Furstenberg's China Club design.

The exquisite artwork hanging from the walls – photographs Diane von Furstenberg took on her travels around Europe, Africa and Asia – is heavily inspired by nature and provides the finishing touches to this luxurious suite.

一直被誉为是英国设计天才的展示柜，克拉里奇邀请世界著名的时尚设计师黛安·冯·芙丝汀宝创作了一系列新室内装饰。大钢琴套房毫不费力地将克拉里奇的光芒与传奇般的DVF优雅风格融合在一起。大理石壁炉、奢华的织物、亲手挑选的家具和标志性印花布，凭借这间有史以来最具时尚风格的套房，黛安将她的时尚标签牢牢的贴在了克拉里奇的历史上。

一提到黛安·冯芙丝汀宝我们就会联想到她标志性的设计裹身裙 (wrap dress)，象征着轻逸优雅与自强不息。而克拉里奇酒店被誉为伦敦装饰艺术的瑰宝，以其繁复精美的装饰久负盛名。DVF的加入让这座艺术瑰宝仿佛穿上了时尚的裹身裙，仍然优雅，但多了些活力。

当客人走入钢琴套房的大门，眼前的第一印象令人惊喜——从宽敞门厅、耀眼的金色色调、黑漆餐桌到定制的豹纹地毯都吸引着客人的目光。黑色的皮面私人吧台像一个魔术盒子般敞开，展露出满满的经典酒品和美味食品。

钢琴套房保留了克拉里奇独特传统魅力，也展现出受到了DVF的影响。原始的克拉里奇家具，包括一架摆在客厅中心的漂亮大钢琴，同时装饰品的效果在慕拉诺玻璃灯的照耀下显得越发地完美。

主卧采用柔和配色方案和高端照明系统，次卧与此相同，次卧在门厅处有一个单独的入口，引入注目的中国会窗帘，明亮的色彩和加粗印花图案让整个房间透露出时尚的感觉——包括一张扶手椅也是DVF中国会的设计作品。

墙上精致的艺术品——DVF环游欧洲、非洲和亚洲时的照片——灵感源自自然，为套房的奢华作出最后的润色。

# Claridge's

克拉里奇酒店

**Completion/Renovation date:** 1898/2011
**Location:** London, UK
**Designer:** David Collins, David Linley, Diane von Furstenberg, Thierry Despont, Dale Chihuly
**Photographer:** Claridge's

**完成/翻新时间**：1898年/2011年
**项目地点**：英国，伦敦
**设计师**：大卫·科林斯，大卫·林内，黛安·冯芙丝汀宝，提埃里·迪邦，戴尔·奇胡利
**摄影师**：图片由克拉里奇酒店提供

Centrally located in London's Mayfair, moments from Bond Street and Hyde Park, Claridge's is the home of timeless English glamour - legendary, illustrious, reassuring and comforting with a rare generosity of spirit and space.

Committed to preserving its unique traditions and providing the latest in creature comfort for its discerning guests, Claridge's recently unveiled the results of an astounding restoration by some of the world's leading designers, including David Linley and Guy Oliver. Bedrooms and suites are individually designed and supremely comfortable assuring guests a truly memorable stay. As a guest in one of the luxurious feature suites, you'll be waited on hand and foot by your personal butler, who will deliver a discreet and professional style of service that is unmistakably English.

There are two penthouse suites on the seventh floor, with outside terraces and striking views across London. These have been magnificently restored: the Brook Penthouse to its former art deco glory, the Davies Penthouse in Edwardian Victorian style. Both provide the ideal retreat in the heart of the city and are ideal for entertaining.

Encounter a truly unforgettable dining experience at Claridge's. Indulge in the decadence and dazzle of the Foyer, where guests can enjoy live musical entertainment every afternoon and evening whilst sampling award-winning afternoon tea or sipping cocktails. Dale Chihuly's magnificent silver-white 'light sculpture' graces this elegantly restored Foyer, perennially popular with Londoners and visitors alike.

The art deco cocoon of the Reading Room, with its contemporary cuisine, offers an intimate and relaxed atmosphere. Featuring marble fireplaces, rich leather columns, banquettes and suede walls, it is the perfect choice for all-day dining.

Alternatively enter a luxurious world of fine dining and indulge in the legendary gastronomy of Gordon Ramsay at Claridge's where you can experience skilfully prepared modern European cuisine.

The seriously stylish Claridge's Bar plays host to many of London's influential set. Designed by David Collins, the original art deco features are beautifully complemented by a silver-leafed ceiling, green glass chandelier and red leather banquettes. Claridge's Bar is renowned for flamboyant cocktails and the finest champagnes, a huge selection awaits the dedicated expert.

The sumptuous Fumoir, steeped in history, will take you back to the fashionable Thirties. Take a seat at the marble horseshoe bar and make your choice from a careful selection of vintage cognacs, armagnacs, rums, tequilas and ports – or sip on a cocktail inspired by the era and served in a Lalique glass.

With its glamorous art deco heritage and timeless elegance, Claridge's offers outstanding facilities for social occasions and all the modern technological equipment to make business events an outstanding success. As a venue for events, Claridge's is unsurpassed in terms of style and prestige, where one can experience stunning design and flawless service, unmatched anywhere else in London.

Tucked away on the 6th floor, discover a haven of tranquillity at Claridge's Health Club and Spa where Sisley products are used for pampering. Enjoy exquisite views across London's rooftops and soothe, recharge and re-energise your mind and body.

位于伦敦梅菲尔街中心,比邻邦德街和海德公园,克拉里奇是永恒英式魅力的代表家园——传奇、辉煌、安心、舒适并略带一丝精神和空间上的宽宏大量。

秉承保留独特传统和为尊贵客人提供物质享受的决心,近来,克拉里奇经由一些世界级的设计师重新翻修,这些设计师包括大卫·林内和盖伊·奥利弗。卧室和套房被分别设计,以确保提供给客人极致舒适的难忘入住体验。当客人身处一间奢华套房,将会享受到私人管家无微不至的服务,管家提供谨慎而专业的无差错英语服务。

7楼的两间阁楼套房有室外露台,在这里能享受到伦敦的魅力景色。两间阁楼都被全面的翻修:布鲁克阁楼呈现了原先装饰艺术风格的辉煌,戴维斯阁楼呈现了爱德华维多利亚时期的风格。两间套房都是城市中心的理想休息寓所和娱乐场所。

克拉里奇提供令人难以忘怀的就餐体验。沉浸在拥有奢靡和梦幻氛围的各楼层,在每天下午和晚上,客人们可以在这里聆听音乐,品尝下午茶或者鸡尾酒。戴尔·奇胡利宏伟的银白色"光雕像"一直 受到伦敦本地人和游客的喜爱,这也使整个翻修后的楼层显得更加优雅。

阅读室酒廊的装饰艺术茧餐厅提供现代菜肴以及亲切,放松的氛围。大理石壁炉、昂贵的皮质柱子、软椅和山羊皮墙面,这一切都验证出这里是全天候就餐的最佳选择,还可以选择进入一个奢华的就餐世界,在这里能品尝到戈登·拉姆齐的传奇烹饪,体验最顶级的欧洲菜品。

克拉里奇的酒吧风格可以让客人体验到纯正的伦敦酒吧场景。酒吧由大卫·科林斯设计,最纯正的装饰艺术特色由漂亮的银色叶子装点的屋顶、绿色水晶吊灯和红色软椅体现出来。克拉里奇酒吧以火焰鸡尾酒和最好的香槟著称,还有很多酒品可供专家级客人品鉴。

华丽的吸烟室,有深深的历史风韵,将带领客人回到时尚的20世纪30年代。坐在马蹄铁型的大理石吧台旁,认真地从法国白兰地、阿马尼亚克酒、朗姆酒、龙舌兰和波尔特酒中挑选中意的一款酒品,或者轻酌一口30年代由Lalique酒杯盛满的鸡尾酒,是客人来到这里的绝佳选择。

带着炫目的装饰艺术和永恒的优雅,克拉里奇提供出众的社交场所需要的设施和现代化的技术设备,确保商务活动的成功举行。作为举办活动的著名场所,克拉里奇的时尚和威望是难以被超越的,在这里,人们可以体验最令人惊叹的设计和无可挑剔的服务,这在伦敦的其他地方是难以可以与之抗衡的酒店。

漫步至6层,可以发现一个宁静的港湾,克拉里奇健身中心和水疗中心提供配套Sisley美容产品的贴身护理服务。在这里,可以一边享受伦敦精致的半空景色,一边安抚、恢复和舒缓身心。

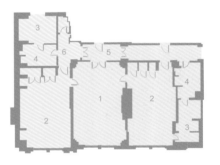

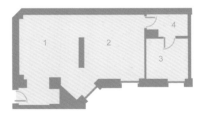
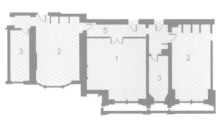

| | |
|---|---|
| 1. Living room | 1. 客厅 |
| 2. Bedroom | 2. 卧室 |
| 3. Bathroom | 3. 浴室 |
| 4. Dressing room | 4. 更衣室 |
| 5. Corridor | 5. 走廊 |
| 6. Guest WC | 6. 客用洗手间 |
| 7. Entrance hall | 7. 门厅 |
| 8. Walk-in wardrobe | 8. 步入式衣橱 |

1. The interior of lobby
2. The magical dining space in the foyer
3. The wedding set-up in ballroom
4. The glamorous and vibrant Claridge's Bar
5. The doorman with guest
6. The bedroom in Grand Piano Suite
7. Deluxe King Room
8. The living room in Grand Piano Suite
9-11. The living room in Linely Suite
12. The bedroom in Linely Suite
13. The bathroom in Linely Suite

1. 酒店大堂内饰
2. 门厅充满梦幻的就餐空间
3. 宴会厅婚礼布置
4. 迷人又充满活力的克拉里奇酒吧
5. 看门人与客人
6. 大钢琴套房的卧室
7. 豪华国王客房
8. 大钢琴套房的客厅
9-11. 林内套房的客厅
12. 林内套房的卧室
13. 林内套房的浴室

# The Renovated Classic by a Fashion Generalist

时尚多面手的创新经典

Karl Lagerfeld's name reverberates like thunder in the fashion world. He has a legendary career. He has successively held the post of brand designer in Jean Patou, Chloe, Fendi and Chanel. He is a versatile person. Besides pushing out his own brand which fully expresses his design individuality, he is also a photographer. Most of the advertisements of Chanel, Fendi and other brands under the same name were photographed by him. Always wearing sunglasses, white long braid, with the expressions as a signboard of Chanel art director, Karl Lagerfeld has been energetically leading the fashion world forward. And this time he expanded his creative realm to the interior design. Alma Schlosshotel im Grunewald is situated among the top grade villas in Grunewald Berlin. The hotel, which is of elegance and splendour, was built in 1912 as a residential palace for Dr. Walther von Pannwitz, a confidant of Kaiser Wilhelm II. It is a challenge for Karl Lagerfeld to renovate this historical building. It is similar to the situation when he undertook the work of Fendi and Chanel, which needs him not only to preserve the style of this district, but also to innovate and break through.

Karl Lagerfeld accomplished his mission perfectly. Through his personalised design and the exquisite renovation of the hotel facilities, the typical Berlin Classicism of 18th century in the hotel is strengthened. Combined with the modern interior design concept, through the exquisite design of details, Karl Lagerfeld refined and highlighted the nature and essence of the hotel, that is, classic or eternal beauty.

卡尔·拉格菲尔德的大名在时尚界真是如雷贯耳。他经历传奇，先后在尚巴度、克洛伊、芬迪和香奈儿担任品牌设计师。他是个多面手，除了推出自己的品牌，淋漓尽致的表现出自己的设计个性，他还是名摄影师，香奈儿、芬迪和同名品牌广告多数由其拍摄。永恒的墨镜、白发长辫，再配以香奈儿艺术总监招牌式的表情，卡尔·拉格菲尔德精力充沛地引领着时装界的前进。而这次他将自己的创作领域扩展到室内设计领域。阿尔玛格吕内瓦尔德皇宫酒店位于柏林格吕内瓦尔德区高档别墅区。酒店建于1912年，原为威廉二世皇帝心腹瓦尔特范潘维茨医生的寝宫，整体典雅优美，富丽堂皇。对这样一座历史建筑进行翻新，这对卡尔·拉格菲尔德提出的挑战与他接手芬迪以及香奈儿时的情形何其相似，既要保持这片地区的气派风格，又要有所创新、突破。卡尔·拉格菲尔德完美地完成了这一使命。通过他的个人设计以及酒店设施的精致翻新，酒店典型的18世纪柏林古典主义被加强。结合现代室内设计理念，透过处处精心的细节设计，卡尔·拉格菲尔德将这座酒店的本质和精华提纯突出，即：经典，或曰永恒之美。

# Alma Schlosshotel im Grunewald

阿尔玛格吕内瓦尔德皇宫酒店

**Completion date:** 2007
**Location:** Berlin, Germany
**Designer:** Karl Lagerfeld
**Size:** 3,600 m²
**Photographer:** Alma Schlosshotel im Grunewald

完成时间：2007年
项目地点：德国，柏林
设计师：卡尔·拉格菲尔德
规模：3600平方米
摄影师：由阿尔玛格吕内瓦尔德皇宫酒店提供

Situated in one of the most exceptional areas of Berlin, the exclusive district of Grunewald, the ALMA Schlosshotel im Grunewald is a benchmark hotel in the German capital. Its elegance and spaciousness, the determined commitment to art and design, and its location within the impressive Grunewald, make it one of the most beautiful hotels in the city.

A sublime mansion redesigned by Karl Lagerfeld and surrounded by picturesque trees and parkland in an exclusive residential district, one of the most historic and plush hotels in Berlin. The Alma Schlosshotel im Grunewald is deliciously stylish, very chic, quietly elegant and lies at the heart of Grunewald – home to hunting lodges and embassies – at the western limit of Charlottenburg. The interiors are breathtaking and have been updated by world-famous Lagerfeld to add an undeniable contemporary feel. The city's central locations can be quickly reached via the motorway – this resort in the city offers pure relaxation and rest far away from any city noise.

It was built in 1912 as a residential palace for Dr. Walther von Pannwitz, a confidant of Kaiser Wilhelm II. The style of that area has been preserved in all of its splendour, an asset which has been enhanced by the personal work undertaken by Karl Lagerfeld, and the exquisite and recent renovation of its facilities.

Reminiscent and typical of the Berlin classicism of the 18th century. The essence and example of the most contemporary interior design. From the tapestries of the carpets, from the woods to the marble and glass, every material, every detail of the Schlosshotel forms part of a whole whose quality is a classical, in other words, timeless beauty.

53 different rooms, each with its own character and personal charm. The Schlosshotel is committed to a limited number of rooms in order to safeguard the privacy of its guests, and to treat them sincerely as part of an exclusive service.

坐落在柏林最特别的地区，唯一的格吕内瓦尔德区，阿尔玛格吕内瓦尔德皇宫酒店是这个城市乃至国家的酒店基准。它的优雅和宽敞，对艺术和设计的坚守，以及它在格吕内瓦尔德的地理位置让这座酒店成为这个城市中最美丽的目的地之一。

卡尔.拉格菲尔德将这座衰败的宫殿重新设计，在这片专用的居住区内，用如画般的树林和风景区将其环绕，成为柏林最有历史意义，最豪华的酒店之一。阿尔玛格吕内瓦尔德皇宫酒店的时尚令人心醉，它别致、淡雅，偎依在格吕内瓦尔德的中心地区，夏洛腾堡的西边。酒店的室内设计是令人惊喜的，由世界著名的时尚设计师卡尔.拉格菲尔德重新设计，使之加上了一种无可争辩的现代氛围。在这里走高速公路可以快速的到达城市中心——这座城市中的度假村远离城市喧嚣，提供让人纯粹放松和休息的机会。

酒店建于1912年，是威廉二世皇帝心腹瓦尔特范潘维茨医生的寝宫。这片地区原有的宏伟被保留了下来，通过卡尔.拉格菲尔德的个人设计以及酒店设施的精致翻新，使酒店的格调被加强。

这里复苏了怀旧经典的18世纪柏林古典主义，是最现代的室内设计的范例和精华的体现，从地毯上的织锦、木材到大理石和玻璃，皇宫酒店的每一种材料，每一处细节都是这个整体的一部分，它的本质是经典的，换句话说，永恒之美。

53间不同的客房，每一间都有自己各自的特色和魅力。从安保和客人的隐私出发，皇宫酒店控制酒店客房的数量，酒店提供给客人真诚的服务，这是专有服务的其中一部分。

1. The study in Kaiser
2-3. The classic and magnificent private lounge
4. The Lounge
5. The luxurious interior
6. The layout of Restaurant Vivaldi
7. The living room in Executive room
8. The mirror in Vivaldi reflecting the brilliant interior
9. A corner near windows in Restaurant Vivaldi
10. The living room in Library
11. The living room in Executive room
12. The sofa in Kaiser
13. The bathroom in Executive room
14. The bedroom in Grand Deluxe room
15-16. The bedroom in Kaiser
17. The desk in Kaiser
18. The bedroom in Grand Deluxe

1. 皇帝客房书房
2、3. 古典而宏伟的私人休息室
4. 休息室
5. 奢华的酒店内饰
6. 维瓦尔第餐厅的布局
7. 行政客房的客厅
8. 维瓦尔第餐厅的一面镜子反射出金碧辉煌的酒店内饰
9. 维瓦尔第餐厅靠窗一角
10. 图书馆客房的客厅
11. 行政客房的客厅
12. 皇帝客房的沙发
13. 行政客房的浴室
14. 大豪华客房的卧室
15、16. 皇帝客房的卧室
17. 皇帝客房书桌
18. 大奢华客房的卧室

| 1. Entrance | 7. Terrace | 1. 入口 | 7. 露台 |
| 2. Lobby | 8. Le Jardin | 2. 大堂 | 8. 后花园 |
| 3. Bar | 9. Restaurant Alter Wintergarten | 3. 酒吧 | 9. 温室餐厅 |
| 4. Piegelzimmer | 10. Restaurant Vivaldi | 4. 镜子客房 | 10. 维瓦尔第餐厅 |
| 5. Musikzimmer | 11. Hofzimmer | 5. 音乐客房 | 11. 庭院客房 |
| 6. Vestibule | 12. Parkzimmer | 6. 门厅 | 12. 公园客房 |

# The Fashion Collection of Dorchester

多尔切斯特的时尚精选

The Beverly Hills Hotel and Bungalows is a member of the Dorchester Collection. This world top fashion hotel group has its own hotel properties in the fashion cities around the world including London, Los Angeles, Milan, and Paris, which makes it closely related to the fashion world. Keeping the design standard of personalised, luxury, and delicate crafts, each Dorchester hotel creates the most elegant and eternal atmosphere of fashion. Therefore, it attracts many visitors of stars and celebrities in the fashion world, including Coco Chanel, Christian Dior, Christian Lacroix, Jean Poul Gaultier, Stella McCartney, and Laudomia Pucci. Dorchester Collession hotels are regarded as "The Second Home" by these celebrities. They are the best places for inspiration where the graceful atmosphere is also in accordance with their tastes. For example, Christian Dior photographed his groundbreaking fashion works in the Athena Plaza Hotel that he often visited. The Beverly Hills Hotel and Bungalows collected in this book is an idol hotel that has been chased for a hundred years with its fascinating fans palace, symbolic banana leaf wallpaper, luxury materials and fashionable colours. The Oscar movie queen Loretta Young, who is also the best dressed female star in the golden age of Hollywood, has appraised the Beverly Hills Hotel and Bungalows as the best place for the fashion show in the filmdom — Oscar. In 2010, the fashion glamour of Dorchester Collection hotels broke out once more in the fashion world. The fashion award set up by the group will elect a fashion design talent each year. He or she can hold an exhibition of his or her own works in any hotel of Dorchester Collection in the meantime of getting 25 thousands pounds' money award. And one of the election standards is that the luxury, exquisite crafts and the fashion sense which are the same to the hotel atmosphere can be expressed in the design.

贝弗利山庄酒店及别墅是多尔切斯特酒店集团的成员酒店之一。这个世界顶级的时尚酒店集团，在全世界的时尚之都，包括伦敦、洛杉矶、米兰和巴黎都有着自己的酒店产业，这让其与时尚界一直有着千丝万缕的联系。多尔切斯特的各家酒店设计秉持着个性、奢华以及精巧的工艺为客人营造出最典雅、永恒的时尚氛围，因此不断吸引着时尚界明星、名流的光顾，这其中包括可可·香奈儿、克里斯汀·迪奥、克里斯汀·拉克鲁瓦、让·保罗·高缇耶、斯黛拉麦卡特尼和劳多米亚·普奇。多尔切斯特酒店被这些名人视为"第二个家"，这里高雅的氛围同他们的时尚品位相一致，也是他们获得创作灵感的地方，例如克里斯汀·迪奥就在他常去的雅典娜广场酒店拍摄其开创性的时装作品。本书选取的贝弗利山庄及别墅酒店更是受到百年追捧的偶像级酒店，迷人的粉色宫殿，标志性的香蕉叶壁纸，奢华的材料和时尚的用色，奥斯卡影后洛丽泰·扬，也是好莱坞黄金时代的最佳着装女星，曾经评价贝弗利山酒店是举办电影界的时尚表演——奥斯卡颁奖礼——的最佳场所。在2010年，多尔切斯特酒店的时尚魅力在时尚界再次爆发，集团创立的时尚大奖将每年选出一位时尚设计师人才，在获得2.5万英镑奖金的同时还可以任意选择一家多尔切斯特酒店举办一次自己作品的展示活动，而其中之一的选择标准就是设计能够体现出如同酒店氛围一样的奢华、精湛工艺和时尚感。

# The Beverly Hills Hotel and Bungalows

贝弗利山庄酒店及别墅

**Completion/Latest renovation date:** 1912/2012
**Location:** Beverly Hill, USA
**Size:** 12 acre
**Designer:** Adam Tihany
**Photographer:** The Beverly Hills Hotel and Bungalows

**完成/翻新时间:** 1912年/2012年
**项目地点:** 美国,贝弗利山庄
**规模:** 12英亩
**设计师:** 亚当·蒂哈尼
**摄影师:** 由贝弗利山庄酒店及别墅提供

As it celebrates 100 years as a second home to Hollywood royalty, local community members and visitors from around the world, Dorchester Collection's The Beverly Hills Hotel and Bungalows is heralding the future with a restoration led by Tihany Design and set to take place in phased stages between now and 2015.

The thoughtfully-crafted restoration will hone and honor the original patina of The Beverly Hills Hotel and Bungalows, restoring and polishing its iconic design elements and enriching its warm residential feel. The hotel lobby, Polo Lounge, guest rooms and suites, and famed pool and cabanas will all be enhanced and refined while respecting their original style, and the hotel's signature red-carpet entrance, C.W. Stock well-designed banana leaf wallpaper, green and white stripes, and famous pink exterior will remain constants. Project work will be conducted in various phases over the next three years so as not to impact the operation and inconvenience guests.

Leading restoration efforts for The Beverly Hills Hotel and Bungalows will be accomplished designer Adam Tihany, widely regarded as one of the world's leading hospitality designers and an inductee in the Interior Design Hall of Fame. The first stage of Tihany's meticulous process is already underway and focuses on the hotel's beloved lobby, where new lighting, carpets and a beautiful banana leaf center medallion crafted of limestone will salute the historic beauty of this popular gathering space. The lobby is scheduled to be complete by July 2012, followed by a gradual enhancement of guest rooms in the hotel's main building.

All guest rooms are expected to be refreshed by early 2014, with 20 to 35 rooms slated to undergo restoration at any one time. Completed accommodations will showcase many sophisticated new features and amenities, including wall mounted B&O TVs with integrated technology in all rooms, illuminated mirrors with built-in TVs, and updated fixtures in a contemporary polished stainless steel finish in the bathrooms. Guests will also appreciate material upgrades in soft leathers, rich mohairs and opulent silks, all designed to add luxury and relevance to the guest experience. Fabric and leather choices in a range of light creams and taupes with silvery blues, greens and yellows accent the rich brown walnut, ebonised oak and parchment lacquered furniture pieces. Antique bronze custom light fixtures throughout and custom wall coverings in the closets and bathrooms were designed exclusively to enhance the feeling of modernity while keeping a chic residential feel.

The legendary Polo Lounge will undergo its own subtle yet carefully-fashioned refresh, further cementing it as the ultimate dining nexus for Los Angeles' most glamorous set and the pinnacle of Hollywood's power dining scene. Then in early 2013, the celebrated pool cabanas and Cabana Café will be upgraded in keeping with the overall design theme.

"The Beverly Hills Hotel and Bungalows will forever be a timeless fixture in the community," said Edward A. Mady, regional director, West Coast, USA and general manager. "We look forward to implementing a restoration which will lovingly highlight the hotel's ties to the spirit and history of Beverly Hills – and which will

provide an effortlessly relaxed new California hotel experience in a one-of-a-kind, legendary setting. Most importantly, we will take great care to maintain much of this beautiful building's originality and landmark features while also staying internationally relevant with the next generation of luxury travellers."

"The essence of the Beverly Hills Hotel and Bungalows is a sparkling cocktail of design, style and Hollywood glamour," says Adam Tihany. "I am enjoying every moment of polishing and updating this gem," he adds.

1. Bungalow Garden
2. Crystal Ballroom
3. Polo Lounge Bar
4. Fountain Coffee Shop
5. The lobby of the hotel
6. Presidential Bungalow Pool
7. The interior of The Bar Nineteen 12
8. The exterior of The Bar Nineteen 12
9. The living room in Beverly Hills Suite
10. Presidential Suite
11. The bedroom in Bungalow
12. The living room in Bungalow
13. The bedroom in Deluxe
14. The bedroom in Presidential Bungalow
15. Premier King with Balcony

1. 别墅花园
2. 水晶宴会厅
3. 马球酒廊
4. 喷泉咖啡厅
5. 酒店大堂
6. 总统别墅游泳池
7. Nineteen 12酒吧内饰
8. Nineteen 12酒吧外部
9. 比佛利套房的客厅
10. 总统套房
11. 别墅客房的卧室
12. 别墅客房的客厅
13. 豪华套房卧室
14. 总统别墅的卧室
15. 带阳台的首相国王客房

作为好莱坞明星、当地社区会员以及游客们的第二个家，贝弗利山庄酒店及别墅迎来了第一个百年庆典。这家多切斯特酒店集团旗下的酒店以重新翻修的方式为自己庆生并向未来宣誓，酒店聘请蒂哈尼设计事务所翻新酒店，工期将从2012年持续至2015年。

精选设计的翻修方案令贝弗利山庄酒店退去锈迹，重新被翻新打磨出原有的殿堂级设计风貌，并被附之温馨的住宅式氛围。在保留原有酒店风格的同时，酒店大堂、保罗休息室、客房及套房，以及酒店著名的游泳池及更衣室都将被维护翻修。在酒店的标志性红毯入口，由 C.W.斯托威尔设计的香蕉叶壁纸、绿白装饰条纹，以及著名的粉色外立面将保持不变。在未来的三年，翻修工程将分不同的工期进行，避免影响到酒店的正常运转或者引起客人的不便。

最重要的翻修工作将由设计师亚当·蒂哈尼指导完成，他被认为是当今世界上最顶级的酒店设计师之一，同时也是室内设计名人堂的入选者之一。蒂哈尼细致翻修工作的第一个阶段是对最受人爱戴的大堂进行翻修，在这里，新的灯光、地毯和香蕉叶壁纸，凸显着用石灰岩制成的浮雕，仿佛在向这个富有历史韵味的聚会空间致意。酒店大堂的翻修在2012年7月全部完成，接下来的工作将是对酒店主建筑中的客房进行维护和翻新。

酒店内所有的客房将在2014年全部被焕然一新，预计将有20到35间客房会被陆续翻修。翻修后的新酒店设施具有许多的新特色和优点，其中包括，在所有的客房内都配有功能全面的壁挂式B&O牌电视，在浴室内配有嵌入电视、照明的镜子和重新被磨光的不锈钢卫浴产品。客人同样会喜欢升级的客房用品，柔软的皮革、密实的马海毛以及奢华的丝绸，这些设计不但增加了酒店的奢华感，还说明酒店将客人的感受放到第一位。织物和皮革的选择在色调上采用了一系列的淡冰淇淋色。此外，灰褐色搭配清脆的蓝色、绿色和黄色凸显了深棕色的胡桃木家具、乌木色的橡木家具和仿羊皮纸涂漆家具。在保持一种俏皮居家感的同时，客房内的衣柜和浴室装饰

着古铜色的定制吊顶，涂着定制墙面漆，这样特别的设计加强了室内设计的现代感。保罗休息室将经历翻新会愈发精致又不失时尚感，让其成为洛杉矶最具魅力，乃至好莱坞最顶级的就餐地点。再接下来迎接翻新的是酒店最著名的游泳池更衣室以及池边的卡巴那咖啡厅，以保持整体酒店设计风格的统一。

"贝弗利山庄酒店将会永远成为这片区域内永恒的珍宝"，爱德华A.麦迪，美国西海岸区域总监兼总经理说。"我们期待一场全面有效的翻修，这将会凸显酒店与贝弗利山庄灵魂及历史的联系——在这样独一无二、富有传奇色彩的场景中，客人们将会轻松地在新式的加州酒店中获得体验的快感。最重要的是，我们将精心的保留酒店建筑的原始面貌及丰碑式的特色，与此同时能够满足全球不同地区年轻尊贵客人的需要。"

"贝弗利山庄酒店及别墅的精髓就像一杯燃烧的鸡尾酒，充满了设计感、时尚感和好莱坞魅力"，设计师亚当·蒂哈尼说到。"我非常享受打磨翻修这颗珍宝的每一个时刻"，他补充到。

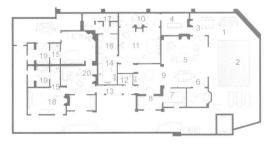
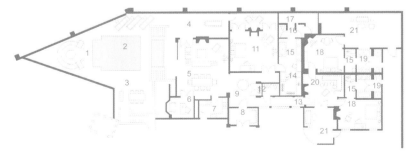

| | | | | | |
|---|---|---|---|---|---|
| 1. Outdoor lounge | 8. Entrance | 15. Vanity | 1. 户外休闲区 | 8. 入口 | 15. 通道 |
| 2. Pool | 9. Foyer | 16. Master bathroom | 2. 游泳池 | 9. 前厅 | 16. 主卧浴室 |
| 3. Outdoor dining area | 10. Private patio | 17. Indoor/Outdoor shower | 3. 户外就餐区 | 10. 私人露台 | 17. 室内/室外淋浴 |
| 4. Outdoor treadmill | 11. Master bedroom | 18. Bedroom | 4. 外跑步机 | 11. 主卧 | 18. 卧室 |
| 5. Great room | 12. Guest powder room | 19. Bathroom | 5. 古丽得客房 | 12. 客用化妆室 | 19. 浴室 |
| 6. Study | 13. Gallery | 20. Living room | 6. 书房 | 13. 画廊 | 20. 客厅 |
| 7. Kitchen | 14. Walk-in closet | 21. Courtyard entrance | 7. 厨房 | 14. 步入式衣橱 | 21. 庭院入口 |

# Hot Destinations of Celebrities
名人们的时尚聚集地

Here are the fashion hotels some stars and social celebrities often visit. In this part, several hot destinations are explored. With steadfast personal faith and pursuit, these fashion designers design fascinating settings for revelry and holiday. The wild and luxurious style UXUA Casa Hotel designed by Wilbert Das, the former creative director, is one of the favourite places appealing prominent figures such as Naomi Campbell; Tortuga Bay, an embodiment of delicate and elegant creative style, designed by Oscar de la Renta, who returned homeland Dominica, is permeated with the charm of royalty; Phillip Treacy who is honoured as "Supreme Hat Magician" made a fashion experiment on the g Hotel in Ireland with different light designs, perfect colour matches and details to attract trendsetters. On the subject of social celebrities and fashion gathered cities, Paris and London are preferential places in which W Paris-Opéra and The May Fair Hotel, London are the representatives of top-fashion taste of each region.

这是一些明星和社会名人经常光顾的时尚酒店。在这部分，我们细致探索几家世界各地的潮店，时尚设计师们秉持各自的个性信念和追求，设计出最吸引人眼球的狂欢、度假场所。吸引潮流人士的眷顾由威尔伯特·达斯（Wilbert Das, 前 Diesel 品牌的创意总监）设计的野奢风格度假村，吸引了娜奥米·坎贝尔等名人的度假首选；回到祖国多米尼加的奥斯卡·德拉伦塔用他一贯的精致典雅创作出龟岛海湾酒店，充满名媛风韵的酒店不乏名流的眷顾；帽子魔术师菲利普·崔西将爱尔兰的 g 酒店当做了一次时尚实践，这里与众不同的灯光设计，色彩搭配以及细节都吸引着潮流人士的光临；说到名人和时尚聚集的城市，自然要谈到巴黎和伦敦，双城中的两所酒店——W 巴黎歌剧院酒店和伦敦梅菲尔酒店能够充分代表各地区的顶级时尚品位。

# Paris Is in the "WOW" Now!

巴黎WOW时尚进行时！

W hotel brand is an exclusive brand which is created by Starwood Hotels & Resorts especially for the trendy people who love fashion. In 1998, the first W brand hotel was opened in New York, where the design keynote for W brand was set up. It is a fashion experience hotel which collects the advanced thoughts in the fields of fashion, art and music, creates inspiration and stimulates sources. Immediately as a guest makes his/her entrance to the hotel, he/she will be surprised to say no words but "WOW". This is just the core glamour that W hotel mostly want to show. The hotel is designed especially for a certain group of trendy people who pursue innovation, fashion and difference. Correspondingly the relevant advanced people in the fields of fashion, music, and art are invited to participate in the overall design of the hotel. The hotel does it meticulously in the crossover cooperation in art circles. Each year W specially holds many kinds of activities relevant to fashion and art, where a new generation of excellent fashion designers and artists will be introduced. The hotel provides a broader space for them, and in the meantime, it can also absorb nutrients of fashion and art, which influences the interior design of the hotel and creates a space of fashion and art that advances with times. W has a chief fashion director in order to keep its eternal fashion style. The present fashion director is Jenné Lombardo who has been keeping close contact with the artists and designers in the aspect of fashion, beauty, and popular culture. He seeks new cooperation and inspiration, explores innovative thinking, and then finds more young designers who can bring vigour of fashion and art to W brand. Now W is cooperated with CFDA (Council of Fashion Designers of America) to hold the "Fashion Incubator" activity, which aims at electing a rising power of fashion designers in the 42 W hotels from all over the world. This opening of W Paris Opera Hotel not only brings the W brand to France, but also creates an opportunity for W fashion culture and Paris fashion culture to conflict and mingle. The exquisite classic architecture of this Paris hotel is endowed with a new life. On the basis of maintaining its original classic building style, it adds an avant background wall, personalised work of art and ornaments, window decorations and creative printing beddings. The particular and exquisite fashion style of Paris is totally shown in each detail of the hotel.

W品牌酒店是喜达屋酒店及度假村集团为热爱时尚的潮流人士打造的专属品牌。1998年，W品牌在纽约开了第一家酒店，从而也为W品牌奠定了设计基调：汇聚时尚、艺术和音乐领域的先锋思想、创造灵感和激发源泉的时尚体验酒店。步入酒店的一刻，酒店会让人惊喜到无话可说，只发出一声"WOW"的感叹，这正是W酒店最想展示出的核心魅力。酒店为一群特定的客人专门设计，他们是一群追求创新、追求时尚、追求与众不同的潮流人士。相应的，酒店的全方位设计也相应的邀请时尚界、音乐界和艺术界的先锋人士一同参与。酒店将跨艺术界的合作做到细致入微，在时尚性和艺术性上，每年W品牌都会专门举办各种时尚和艺术类的相关活动，推荐出新一代的优秀时尚设计师和艺术家，在为他们提供更广阔的平台的同时，也会不断的吸取时尚与艺术的养分，从而影响着酒店的室内设计，打造出与时俱进的时尚艺术空间。W品牌为了保持永恒的时尚风格设有专门的时尚总监，目前的时尚总监杰内·隆巴尔多（Jenné Lombardo）一直与在时尚、美容和流行文化方面的艺术家和设计师们保持着紧密的联系，寻求新的合作和灵感，发掘创新思维，以此发现更多的年轻设计师为W品牌注入时尚艺术的活力。目前，W品牌就与CFDA (Council of Fashion Designers of America 美国时尚设计师协会）合作举办"时尚保温箱"（Fashion Incubator）活动，在全世界的42家W酒店选拔新兴的时尚设计师力量。此次W巴黎歌剧院酒店的开业，不但将W品牌势力带入法国，更是创造出一次将W时尚文化和巴黎时尚文化两者相碰撞融合的机会。酒店将巴黎的精致古典建筑赋予新的生命力，在保持原有建筑古典风韵的基础之上，加之前卫的背景墙，个性的艺术品和装饰品，窗饰雕花和创意印花床上用品，各处细节尽显巴黎独有的精致时尚风韵。

# W Paris - Opéra

W巴黎歌剧院酒店

**Completion date:** 2012
**Location:** Paris, France
**Designer:** W Global Brand Design & Rockwell Group Europe (RGe) – Director: Diego Gronda
**Area:** 86,200 m²
**Photographer:** Starwood

完成时间：2012年
项目地点：法国，巴黎
设计师：W酒店全球品牌设计公司，洛克威尔设计集团欧洲设计总监圣地亚·哥格隆达
规模：86200平方米
摄影师：由喜达屋酒店集团提供

Starwood Hotels & Resorts Worldwide, Inc. (NYSE: HOT) announced the debut of W Hotels Worldwide in France with the opening of W Paris - Opéra. Owned by Barcelona-based Meridia Capital, the hotel is set inside a historic 1870s Haussman-era heritage building near Opéra Garnier, Galeries Lafayette and Place Vendome in the eclectic, historic 9th Arrondissement in Paris. Through W Hotels' unique and distinctive programming, which is inspired by the brand's roots in New York where it was founded in 1998, W Paris - Opéra will set a new scene in the city for both guests and Parisians alike. The opening of W Paris - Opéra marks a significant milestone in the W brand's global expansion as it gears to enter Singapore, Thailand and China later this year.

"Taking the W brand into Paris, a global fashion capital, marks a true milestone in W's global expansion into the world's most exciting and vibrant destinations", says Eva Ziegler, Global Brand Leader, W Hotels Worldwide and Le Méridien. "Fashionable global jetsetters can now be on the runway with W from New York to London and Paris. Unique lifestyle programming within a cutting-edge, designed environment will bring a new, distinctive experience to Paris and the Parisians alike."

W Paris - Opéra provides a cutting-edge lifestyle experience, featuring 91 stylish guest rooms, including 20 suites and two Extreme WOW Suites (W's interpretation of the Presidential Suite). The hotel's signature restaurant, Arola, marks the debut of Michelin-starred chef, Sergi Arola in France. The hotel offers spectacular views of the Opéra Garnier from its ground level and mezzanine floors, including W Lounge, W's take on the traditional lobby, a buzzing social space where visitors can sip, drink and flirt around cocktails. W Paris - Opéra features a SWEAT® fitness centre, as well as stylish meeting and event spaces. Guests can expect the brand's signature Whatever/Whenever® service philosophy, providing guests whatever they want – from the latest in fashion, design, and music in Paris, to the city's most exclusive events and nightspots.

### Design – "The Spark" Inspired by New York's Energy and the Sophisticated Radiance of Paris

Designed jointly by W Global Brand Design and the award-winning Rockwell Group Europe, W Paris - Opéra was conceptualised to create a dynamic dialogue between the building's historic façade, Paris as the "City of Light" and W Hotels' DNA by infusing New York's dynamism and never-ending energy into the quintessentially sophisticated Parisian neighbourhood. GCA Architects, a renowned architecture and interior design firm, also advised Meridia Capital in this project.

### Arola – Bringing Spanish Innovation to Paris

A renowned Spanish Michelin-starred chef from Barcelona, Sergi Arola has been appointed as the executive chef at W Paris - Opéra's signature restaurant, Arola. The restaurant introduces the innovative "Pica Pica" culinary concept of creative dishes and tapas, seasonal products and French classics, served on a platter for all to share. Regarded as one of the most creative minds in Spanish haute cuisine, the Catalan chef has also designed the menu for the hotel's W Lounge and in-room dining

## W Lounge – To See and Be Seen

Set on the ground floor, the W Lounge is set to attract Parisians and jetsetters with its private entrance and wide bay windows looking out to the bustling streets. Guests can enjoy a wide range of tapas and signature cocktails offered by the hotel's mixologist.

Rockwell Group Europe has teamed up with Starwood Hotels & Resorts Worldwide to design the W Paris-Opéra, W Hotels' debut in France. The hotel occupies an elegant 1870s Haussman-era building located across the street

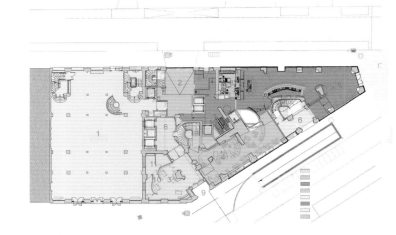

1. The detail in guestroom
2. W Lounge with artwork
3. Arola Restaurant
4–5. The details in living room
6. W Lounge Bar
7. The details in W Lounge
8. Welcome desk
9. Boardroom setup in studio
10. E WOW Suite
11. The living room in E WOW Suite
12. Fabulous Suite
13. Spectacular Room
14. A corner in Spectacular Room
15. The detail in living room of E WOW Suite

1. 客房细节
2. 装饰着艺术作品的W酒廊
3. 阿罗拉餐厅
4、5. 客厅细部
6. W酒廊
7. W酒廊细节
8. 接待台
9. 会议室里的董事会会议布置
10. E WOW套房
11. E WOW套房客厅
12. 美好套房
13. 华丽套房
14. 华丽套房一角
15. E WOW套房客厅的细节

1. **Retail**
2. **Deliveries**
3. **Welcome desk**
4. **Living room**
5. **Bar**
6. **Foyer**
7. **Staff access**
8. **Kitchen**
9. **Entrance**
10. **Safe room**
11. **Back office**

1. 消费区
2. 货梯
3. 接待台
4. 休息大厅
5. 酒吧
6. 前厅
7. 员工通道
8. 厨房
9. 入口
10. 保安室
11. 机房

from the famous Opéra Garnier. Taking the inspiration from combining New York's energy, where the W brand began in 1998, and Paris, the "City of Light," Rockwell Group Europe is flooding the new W Hotel with illumination and colour.

Historic elements throughout the building such as ornamented columns, vaulted ceilings and decorated doors are redefined to create a dialogue and contrast between the old world and the contemporary elements. And an oversized undulating wall of light is the central design feature that defines the building from the inside and from the street as it weaves through the public spaces, secret corners of DJ booths and martini bars, through the corridors and finally into each guest room. This programmable wall of light can transform based on season, event or time of day to create a festive and celebratory atmosphere, bringing the historic building to life with sophistication and a glowing vibrancy.

随着W巴黎歌剧院酒店的开幕，喜达屋酒店度假村集团正式宣布旗下的品牌W酒店进军法国。由总部设在巴塞罗那的美里迪亚投资公司管理，酒店被设置在一栋18世纪70年代的奥斯曼时期的遗产建筑中，邻近巴黎歌剧院、老佛爷百货公司和旺多姆广场，在折衷主义，充满历史感的巴黎第九区。W酒店品牌自从1998年在美国成立，品牌精神也随之根植成长，W巴黎歌剧院酒店正是通过这种独特、与众不同的设计理念在城市中为游客和巴黎人共同设立起新场景。W巴黎歌剧院的开幕标志着重要丰碑的奠立，W品牌的全球扩张计划将在今年年底直指新加坡、泰国和中国。

"W品牌入驻巴黎这个全球时尚之都，标志着W的全球扩张计划在世界最使人疯狂的活力城市中立下了真正的里程碑"，伊娃·齐格勒，W全球酒店及艾美酒店全球品牌领导人评价酒店时说。"全球时尚旅行者现在可以跟随W酒店从纽约到达伦敦和巴黎了。独特、前卫的生活方式，充满设计感的空间环境将带给巴黎和巴黎人一种新式、独特的体验。"

W巴黎歌剧院酒店提供一种前卫的生活体验，内设91间时尚客房，包括20间套房和两间极致WOW套房（W酒店的总统套房）。酒店的标志性餐厅——阿罗拉，标志着米其林主厨赛吉·阿罗拉在法国的第一家餐厅的成立。在酒店底层，可以欣赏到巴黎歌剧院的壮观景色，在夹层，包括W酒廊，W酒店的传统大堂，这里是忙碌的社交场所，在这里客人可以斟饮品尝各式鸡尾酒。W歌剧院酒店配备SWEAT健身中心和时尚的会议活动空间。客人可期待W品牌标志性的Whatever/Whenever®服务哲学，提供给客人一切所想，从巴黎最新的时尚、设计、音乐，到城市最独有的活动和夜总会场所。

**设计——源自纽约的能量和巴黎的精致光辉碰撞出的"火花"**

酒店的设计由W全球品牌设计公司和获奖设计公司洛克威尔集团欧洲公司共同合力完成，W歌剧院酒店营造了一场充满活力的对话，在有"光之城"之称的巴黎，一座有着历史风貌的建筑与融入了纽约活力的W品牌酒店的DNA进行了这次对话，它让无尽的能量注入了标准的精致巴黎生活。著名的建筑及室内设计公司，GCA建筑公司也参与了美里迪亚投资公司的这个项目设计。

**阿罗拉——将西班牙的创新带到巴黎**

著名的西班牙米其林星级厨师，来自巴塞罗那的赛吉阿罗拉担任W巴黎歌剧院酒店的这座标志餐厅——阿罗拉的主厨。餐厅引入了创新的"喜鹊"烹饪理念，提供创新菜式和餐前小吃，应季食材和法国经典菜式。作为西班牙高级烹饪最创造力的厨师之一，加泰罗尼亚厨师同样为酒店的W酒廊和客房餐厅提供服务。

**W酒廊——见与被见**

坐落在酒店一层，W酒廊拥有私人入口和宽敞的飘窗，在这里可以欣赏街道喧嚣的场景，吸引了不少巴黎人和旅行者的目光。客人们可以享受餐单丰富的餐前小吃和酒店调酒师提供的标志性鸡尾酒。

洛克威尔集团欧洲公司与喜达屋酒店及度假村集团合作，设计了W巴黎歌剧院酒店，这是W酒店在法国的第一家酒店。酒店位于18世纪70年代的豪斯曼时代的一座建筑之中，街的对面就是著名的巴黎歌剧院。酒店设计吸取纽约的能量精华，这里是1998年W品牌的诞生地，如今在这座"光之城"——巴黎，洛克威尔将纽约的W品质加入了光亮和色彩。

在整个建筑中，历史元素贯穿始终，例如装饰柱、拱顶和装饰门，这些在酒店中被重新定义，营造出的是一场旧世界与当代元素的对话与冲突。超大的波浪式的光墙是整个酒店的中心设计特色，光从建筑的内部而来，也从酒店外的街道而来，仿佛在酒店的公共区域，DJ亭的神秘交流和马提尼酒吧波动，它穿过走廊，最后进入每一间客房。这种精心设计的光墙可以根据季节、场合、时间改变形式，以此制造出节日的庆祝气氛，带给这座历史的建筑以精致活力的新生。

# The Fashion Growing in the Wild

让时尚在自然中延续

The hotel designed by Wilbert Das, who was the creative director of Italian brand Diesel before, was opened in Trancoso. Wilbert Das came from the south of Netherlands. He was recruited into Diesel once he was graduated in Lanham College of Art in 1988, and was appointed as the creative director 5 years later. He had gradually built Diesel into a brand empire with more than 10 years. The style of Diesel is young and creative, whose inspirations are all from the dribs and drabs of daily life which are closely linked to the trend. The characteristic dirty jeans, which are full of holes and blots, and manually rinsed to be old, have become the sexiest symbol. It may be said that UXUA Hotel is the extension of Wilbert Das's fashion design career. The 4 old Dourado villas and other 6 bungalows are spread in the garden full of fruit trees and green coconut bushes. UXUA Hotel is constructed by local craftsmen with traditional techniques using recycled or local materials for two years. The design concept of Wilbert Das is to combine the local handcrafts with modern space and lighting design concept, which is in accordance with his concept in the fashion field.

意大利品牌迪赛（DIESEL）前创意总监——威尔伯特·达斯设计的酒店在捕鱼小镇托兰克索开业。威尔伯特·达斯出生于荷兰南部。1988年毕业于拉纳姆艺术学院并进入迪赛，五年后威尔伯特·达斯出任迪赛创意总监。十多年来他为迪赛逐渐成为一个品牌帝国立下了汗马功劳。迪赛的风格是青春并且富有创造力，他从日常生活的点滴中汲取灵感并紧随时尚潮流而动。手工漂洗做旧、污渍、满是破洞的牛仔裤是迪赛标志性的设计，这也成为现在最性感的象征。UXUA酒店可谓威尔伯特·达斯时尚设计生涯的延续。UXUA酒店的4间古老瓜德拉多别墅和其他六间小屋散布在种满果树的花园内。椰树灌木，郁郁葱葱。UXUA酒店由当地工匠以传统技法，使用回收或当代材料，耗时两年建成。威尔伯特·达斯的设计理念是将本地的手工艺与当代空间、灯光设计的概念相结合，这与威尔伯特·达斯在时尚领域的理念一脉相承。

# UXUA Casa Hotel

UXUA之家酒店

**Completion date:** 2009
**Location:** Trancoso, Brazil
**Designer:** Wilbert Das
**Size:** 5,000 m²
**Photographer:** @UXUA Casa Hotel

完成时间：2009年
项目地点：巴西·托兰克索
设计师：威尔伯特·达斯
规模：5000平方米
摄影师：由UXUA之家酒店提供

In 2009 Wilbert Das, former Creative Director of Italian fashion label Diesel, opened his eco-chic retreat, the UXUA Casa Hotel (pronounced oo-SHOO-ahh), located in the remote fishing village of Trancoso on Brazil's breathtaking Bahian coast.

Founded by Jesuits in the 16th century, Trancoso has the air of time stood still, preserving over centuries its stunningly simple architecture, unspoiled nature, endless stretches of pristine beaches, and rich local traditions.

Trancoso's heart is its picturesque Quadrado – a grassy hilltop square closed to traffic and anchored at one end by the simple white, 400-year-old São João Batista church overlooking the sea. The Quadrado is surrounded by dozens of brightly painted one storey fisherman casas which have remained unchanged over centuries.

UXUA is composed of four of these antique Quadrado casas, and another six homes (including a rustic treehouse of reclaimed wood) discreetly spread under fruit trees of an adjoining 5,000-square-metre garden. All are filled with Bahian art, antiques, and custom-made furniture. The hotel blends harmoniously into its environment without even signage to indicate its location.

UXUA's lush grounds include a restaurant, bar, library, gym, spa, as well as a stunning pool made of 40,000 pieces of green aventurine quartz, a local crystal thought to possess healing powers; all is complemented by the UXUA Praia Bar, a spectacular beach-front lounge shaped from old fishing boats an 8-minute walk from the Quadrado.

The hotel was built by local artisans working over two years using traditional building methods and reclaimed and organic materials. The project combines local craft with contemporary concepts of space and light which nod to Wilbert's international work in fashion and interior design, with the end aesthetic best labeled rustic modernism. The overall UXUA philosophy is something of the reverse of globalism, but instead a celebration of all things native, an ideal Wilbert Das refers to as localism.

UXUA's emphasis on the local extends to its menu, which is regional Bahian composed of the freshest locally-sourced seafood and produce, as well service, which is deeply personalised, with every detail thought of and delivered by a staff who host in the friendly style of the region, as if welcoming guests into their own home.

The hotel has quickly become a destination of choice for those from the fashion and entertainment fields, being used for prestigious photo-shoots such as the 2010 Pirelli Calendar shot by Terry Richardson, as well as its acclaimed efforts in promoting preservation of local culture and the environment.

2009年，意大利时尚标签迪赛的前创意总监威尔伯特·达斯，创立了他的生态时尚世外桃源——UXUA之家酒店，它坐落在巴西叹为观止的巴伊亚洲海岸——托兰克索的一个偏远的渔村。

托兰克索由16世纪当地的耶稣会建立，在这里依然能够感受到那个时代的空气，这里留存了几个世纪以来令人惊叹的简朴建筑，未被破坏的自然景观，无尽伸展的原始海滩和丰富的当地传统。

托兰克索的中心是它如画般的瓜德拉多———片被青草覆盖的山底广场，这里交通不便，在一头是一个有400年历史的建筑，白色的圣若昂巴蒂斯塔教堂遥望着整片大海。在瓜德拉多周围是一座座涂着明亮颜色的一层渔夫别墅，保留了几个世纪的模样。

UXUA由4间这样的古老瓜德拉多别墅，和其他六间小屋组成（包括一间用回收木料做成的乡村树屋），他们遍布在种满水果树的5000平方米的花园内。这里充满了巴伊亚艺术品、古董和定制家具。在这里甚至没有任何标志显示酒店的地理位置，酒店和谐地与本土环境融合在一起。

UXUA苍翠繁茂的土地包括餐厅、酒吧、图书馆、健身中心和SPA，以及一个非凡的游泳池，游泳池由4000块绿色砂金石石英制成，这是一种当地的水晶，被认为能够恢复能量；普拉亚酒吧使整个酒店充实了起来，这是一间壮观

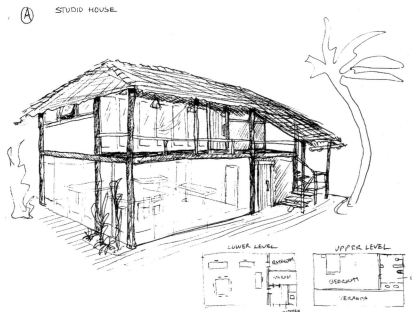

的海滩边休息室，它型如一艘老渔船，距离瓜德拉8分钟的路程。

酒店由当地的艺术工匠经过两年时间，用原始的建筑方式，利用回收材料和有机材料建成。项目结合了当地技艺与当代空间和灯光的设计概念，这正是威尔伯特的时尚作品和室内作品想要表达的，最后，你会发现在酒店的设计中还融入了乡村现代主义的美学价值观。整个UXUA的哲学是某种全球主义的倒退，然而，这并不是单纯表达对任何天然事物的敬意，而应是一种更为理性的叫法，威尔伯特·达斯本人称它为"地方主义"。

UXUA崇尚用当代食材做出的美食，地道的巴伊亚美食包括当地最新鲜的海鲜和海鲜制品，同样的，服务也是个性定制的，每一个细节都经过深思熟虑，再以当地最友好的方式传递给客人，仿佛是欢迎客人来到他们自己的家一样。

酒店很快成为了时尚界和娱乐圈客人们的目的地之选。在特里·理查德森拍摄的2010倍耐力年历中，酒店荣誉入选，同时，酒店对呼吁保护当地文化和环境上作出的贡献也得到了凸显。

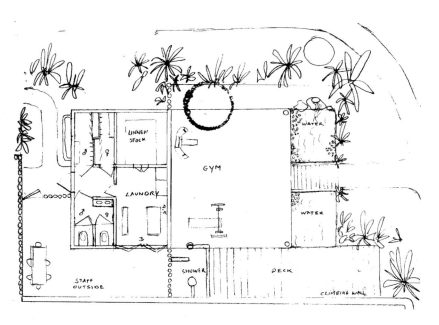
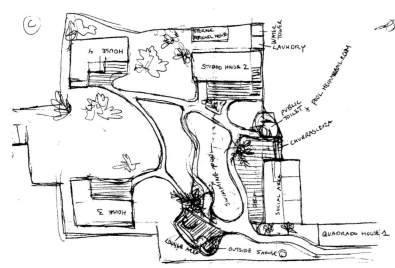

1. The cottage and abandoned fishing boat at UXUA Beach Lounge
2. The overview of UXUA
3. Nozinho casa
4. The UXUA Beach Lounge – UXUA Praia Bar
5. The façade of Seu Irenio casa
6. The path in the garden
7. The relaxation pavilion
8. The living room of Seu Irenio casa
9. The detail of bathroom in Estudio casa
10. The detail in restaurant
11. The Lounge
12. The bathroom of Seu Irenio casa
13. The study in Quintal da Gloria casa
14. The relax room in Seu Joao casa
15. A corner in the restaurant
16. The living room in Nozinho casa
17. The living room in Seu Pedrinho casa
18. The bedroom in Seu Joao casa
19. The bedroom in Estudio casa
20. The layout of bedroom
21. The details in a bathroom
22-23. The details of restaurant

1. UXUA海滩酒吧的废弃渔船草棚
2. UXUA 度假村全景
3. 诺吉奥别墅屋
4. UXUA海滩酒吧—UXUA普拉亚酒吧
5. 秀伊诺别墅外部立面
6. 花园小径
7. 放松凉亭
8. 秀伊诺别墅客厅
9. 书房别墅的浴室
10. 餐厅细节
11. 休息区
12. 秀伊诺别墅的客厅
13. 歌莉娅别墅的书房
14. 秀热奥别墅的卧室
15. 餐厅的一角
16. 诺吉奥别墅的客厅
17. 秀林奥别墅的客厅
18. 秀热奥别墅的卧室
19. 书房别墅的卧室
20. 卧室布局
21. 餐厅细节
22、23. 一间浴室的细节

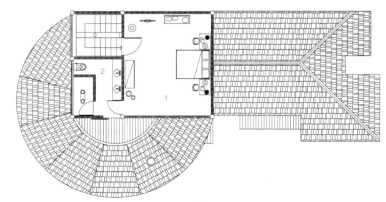

1. Suite
2. Bathroom

1. 套房
2. 浴室

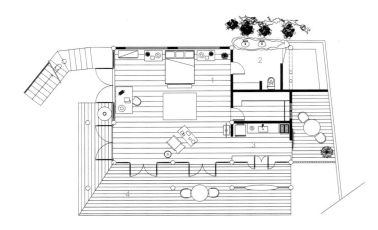

| 1. Suite | 4. Living room | 1. 套房 | 4. 客厅 |
| 2. Bathroom | 5. Outdoor kitchen | 2. 浴室 | 5. 户外厨房 |
| 3. Kitchen | 6. Terrace | 3. 厨房 | 6. 露台 |

| 1. Suite | 3. Kitchen | 1. 套房 | 3. 厨房 |
| 2. Bathroom | 4. Balcony | 2. 浴室 | 4. 阳台 |

# Dominican Socialite Holiday Feeling

多米尼加的名媛度假风情

Oscar de la Renta is one of the America's most iconic fashion designers having built a great reputation particularly in haute couture, red carpet gowns and evening wear. The award-winning designer's fashion house has been dressing leading international figures, film stars and royalty since the 1960s and continues today. Crafting clothing that women desire, Oscar de la Renta is skilled in creating designs that are simple and delicate yet dramatic and luxurious all at once. The result is nothing short of elegant.

Born in the Dominican Republic, Oscar de la Renta aimed to create the interiors of Tortuga Bay to emphasize the privacy and sense of calm that he loves and cherishes in the Dominican Republic. The hotel's villas feel much like de la Renta's designs: gorgeous, exquisite, and elegant. Facing a stretch of private beach along the Caribbean Sea, the lush plants, white sand, blue water and soft breeze create an atmosphere of peace and quiet throughout the villas. Ranging in size from one-bedroom to four-bedroom, each villa includes its own lounge area and outdoor terrace. Bedrooms on the second floor have exclusive balconies and magnificent views of the sea with room furnishings made by local artisans.

Oscar de la Renta's elegant design brings the exotic charm of the Dominican Republic to life at Tortuga Bay.

奥斯卡·德拉伦塔是美国最受崇敬的时尚设计师之一,在高级时装,红毯礼服和晚礼服设计领域久负盛名。从上世纪60年的至今,这位获奖设计师深受国际名人,电影明星和皇室的喜爱。他设计的礼服典雅华丽,做工考究,极具质感,德拉伦塔深谙追求时尚的名媛需要,擅长创作简洁精致,充满戏剧性又不失华贵的风格,尽显优雅。

出生于多米尼加的奥斯卡-德拉伦塔在设计这家完全由别墅组成的酒店之初,就将它的风格定位在既具有浓郁的多米尼加风情,强调私密性和宁静的感受。整个酒店给人的感觉就像奥斯卡-德拉伦塔的女装一样:华丽、精致、典雅。别墅面向加勒比海,脚边是长约两英里的私人海滩。葱绿的植物,洁白的沙滩,碧蓝的海水,精致的别墅,轻柔的风营造出平和宁静的氛围。别墅从一间到四间客房大小不一,所有客房都配有躺椅区和露台,二楼的卧室拥有宽敞的专属阳台和壮丽的海景。屋内的家具都由当地工匠制作,保证原汁原味。

德拉伦塔的优雅设计将魅力国度的异域风情融入到客人们的假日生活之中。

# Tortuga Bay

龟岛海湾酒店

**Completion/Latest renovation date:** 2006/2011
**Location:** Punta Cana, Dominican Republic
**Designer:** Oscar de la Renta
**Size:** 3943.87 m²
**Photographer:** PUNTACANA Resort & Club

**完成/翻新时间：** 2006年/2011年
**项目地点：** 多米尼加共和国，篷塔卡那
**设计师：** 奥斯卡·德拉伦塔
**规模：** 3943.87平方米
**摄影师：** 由篷塔卡那度假村与俱乐部集团提供

Tortuga Bay, designed by the renowned Oscar de la Renta, is a member of Leading Hotels of the World and the only AAA Five Diamond Awarded hotel in the Dominican Republic and is located within the exclusive PUNTACANA Resort & Club, situated on the eastern tip of the Dominican Republic. Located in a private enclave of the 15,000 acre resort, where de la Renta, Mikhail Baryshnikov and Julio Iglesias keep homes, Tortuga Bay includes 13 luxurious beachfront villas offering unrivalled luxury, privacy, security and five-star service for people who expect the exceptional.

All one-, two-, three- and four-bedroom villas have been designed by de la Renta and are perched along a private white sand beach with every convenience at arm's length. His influence is centred around simplicity, open spaces and beautifully crafted furniture. With respect to the raw beauty of the resort's setting, the villas are brought to life with a palette of natural island colours, along with an authentic combination of wicker and wood furnishings and luxurious linens to create a look that is at once luminous, serene and utterly elegant.

Delivering the level of service the resort's most exclusive guests have come to expect, Villa Managers are always on hand to assist guests of Tortuga Bay with a wide variety of services. Guests are also each provided with a private golf cart for exploring the breathtaking grounds of the property. Additionally, those staying at Tortuga Bay are greeted as VIPs at Punta Cana International Airport (owned and operated by the resort's developers Group PUNTACANA®), where they are met upon arrival and swiftly escorted through both immigration and customs, and then whisked to the resort, just a few minutes away.

Guests of Tortuga Bay have complete access to the first-class amenities available at PUNTACANA Resort & Club, including three miles of white sand beaches, 45-holes of championship golf, eight restaurants, a Six Senses Spa, and a 1,500 acre ecological reserve. In December 2011, Oscar de la Renta opened a new boutique in Tortuga Bay that carries a wide range of items, from clothing to accessories, produced by the high-end fashion designer.

龟岛海湾酒店，由享负盛誉的时尚设计师奥斯卡·德拉伦塔设计，是世界顶级酒店组织的会员之一，也是多米尼加共和国唯一获AAA五钻奖的酒店，酒店位于篷特卡那度假村与俱乐部之内，多米尼加的东端。龟岛海湾位于的区域是一个15000公顷的私人领地，这里也是德拉伦塔、米哈伊尔·巴雷什尼科夫、胡利奥·伊格莱西亚斯的家，酒店拥有13间奢华海滩别墅，给那些期待非常体验的客人们与世无争的奢华，以及私人、安全的五星级服务享受。

所有的单间、双间、三间和四间卧室别墅均由德拉伦塔亲自设计，它们依居在私人的白色海滩旁，离海非常的近。酒店内部风格质朴，空间开放，家具精美。度假村布局之美体现在，别墅将自然小岛的色彩带入客人的生活，在室内，最能体现当地特色的组合——柳条和木材制作的家具搭配奢华的床用织物，营造出夺目、宁静，又极致优雅的氛围。

度假村提供给尊贵客人们的服务是最值得期待的，别墅经理们随时待命，为龟岛海湾的客人们提供服务，满足他们的多项要求。酒店还提供私人高尔夫球车以及供客人欣赏叹为观止的风景。另外，在龟岛海湾居住的客人会被作为篷塔卡那航空公司的会员（由度假村的发展商篷塔卡那集团拥有并管理），一旦客人乘机达到，酒店会即刻护送，疾驰至度假村只需要仅仅几分钟的时间。

3

龟岛海湾酒店的客人在这里还可以完全体验到最先进的篷塔卡那度假村与俱乐部的高端设备，包括三英里的白沙海滩，45洞冠军高尔夫球场，8家餐厅，第六感水疗中心，以及1500公顷的生态保留地。2011年12月，奥斯卡·德拉伦塔在龟岛海湾酒店开设了一家新的精品店，货品一应俱全，从成衣到首饰，均出自这位高端设计师之手。

1. **Bedroom**
2. **Balcony**
3. **Kitchen**
4. **Bathroom**
5. **Walk-in closet**

1. **卧室**
2. **阳台**
3. **厨房**
4. **浴室**
5. **步入式衣橱**

1. The overview of villa exterior
2. Rows of villas
3. The lounge in terrace
4. The facade of villa
5. Pool
6. The lounge inside the villa
7. The details of landscape
8. Whirl pool
9. The signpost for spa in the hotel
10. Couple bath in the spa
11. The lobby in the villa
12. The bedroom in Yellow Room
13. The suite
14. Junior suite
15. Double room

1. 别墅外观全景
2. 成排的别墅
3. 楼台上的休息区
4. 别墅立面
5. 游泳池
6. 别墅内的休息区
7. 景观细节
8. 按摩游泳池
9. 酒店水疗中心标志牌
10. 水疗中心的双人理疗室
11. 别墅内客厅
12. 黄色客房卧室
13. 套房
14. 标准套房
15. 双人房

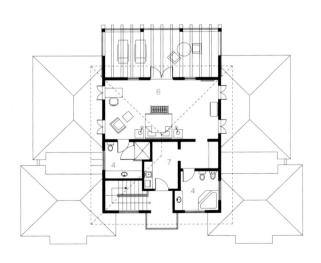
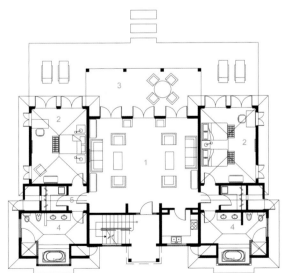

| | | | |
|---|---|---|---|
| 1. Living room | 5. Walk-in closet | 1. 客厅 | 5. 步入式衣橱 |
| 2. Bedroom | 6. Master bedroom | 2. 卧室 | 6. 主卧室 |
| 3. Terrace | 7. Hall | 3. 露台 | 7. 大厅 |
| 4. Bathroom | | 4. 浴室 | |

# The Artistic Practice in Architecture by Hat Magician

帽子魔术师的空间艺术实践

Phillip Treacy is honoured as "Supreme Hat Magician in the World". He, whose talent is focused on the hatwear design, has the same unrestrained and vigorous passion in creation as the designer Alexander McQueen who is also pursued by the famous fashion commentator Isabella Bobrow. Treacy's works had been pursued by many celebrities such as Victoria Beckham, Madonna, Kate Moss, and Naomi Campbell. In the meantime, many top fashion brands had extended the olive brand to him for cooperation. The cooperation with Chanel in 1991 made his name become popular in the fashion world. Nowadays, Phillip Treacy has 20 years of professional brilliance with his heart of art full of creation. In his hatwear works, many kinds of feathers are used as the main materials which can show the grace. There are also some works of futurism and super-realism created by metal gauze, metal plate, wood, and even crystal glass. In these works, it can be seen that what Treacy designed is not only a hat, but also a work of art or even architecture. This makes his viewing angle surpass the original hatwear design. Maybe it was a chance to make him realise that space design can make his art expectations come true more specifically, when he accepted the work of designing the interior of g hotel. In this hotel of Treacy's hometown, he used colours boldly and creatively with gold and purple strongly dotted. In blue lounge and pink salon that are named after colour, these colours are sprinkled in each corner of the space, which feels creative and fashionable. Furthermore, Phillip Treacy's originality and talent can be felt in each corner of the hotel. Treacy's pioneer design thought is shown through the sea horse water tank and reception walls in the reception area, the lighting design of the spa centre, etc. Treacy's ability to use materials creatively is performed perfectly in this hotel. Building stones, timbers, linens and even glass are given a new life. This kind of varied originality can make guests feel the sense of fashion in hotel all the time.

菲利普·崔西被誉为"全世界首屈一指的帽子魔术师",他与同样受到著名时尚评论人伊莎贝拉·布罗追捧的设计师亚历山大·麦昆一样具有天马行空的创作激情,只不过他是将他的天赋施展在帽饰设计上。崔西作品受到过很多名流的追捧,例如维多利亚·贝克汉姆、麦当娜、凯特·莫斯、娜奥米·坎贝尔,与此同时,很多顶级时尚品牌也抛出合作的橄榄枝,与香奈儿在1991年的合作令菲利普·崔西的名字在时尚界迅速崛起。如今的菲利普·崔西经历了20年的职业辉煌,凭借的是其一颗充满创新的艺术之心。在他的帽饰作品中,有用各种羽毛为材料主角的柔美,也有用金属纱网、铁片、木材,甚至水晶玻璃创造出的未来主义和超现实主义作品。从这些作品中,可以看出崔西设计的不仅仅是帽子,而是艺术品,甚至是建筑物。这本身让崔西的视角开始超越原来基本的帽饰设计,当接收g酒店的室内设计时,也许正是这次机会让他意识到用空间设计更能具体的实现他的艺术期望。在这家崔西家乡的酒店,他将大胆的颜色创新地运用,金色和紫色强势地点缀着;蓝色酒廊和粉色沙龙,这些以颜色命名的空间中近乎夸张地在每个角落将这些颜色挥洒,显得非常有创意和时尚感。此外,酒店的各个角落都会令人感受到菲利普·崔西非凡的创意和才能,接待区时尚的海马水缸和接待墙,酒店水疗中心的照明设计等彰显着崔西的先锋设计思想。在酒店中,崔西创意运用材料的能力再次得到完美的发挥,石料、木料、亚麻甚至是玻璃被赋予重新的生命力,这样多变的创意令客人无时无刻体验着酒店的时尚气息。

# The g Hotel

g酒店

**Completion date:** 2005
**Location:** Galway, Ireland
**Designer:** Philip Treacy
**Size:** 101 guestrooms
**Photographer:** The g Hotel

完成时间：2005年
项目地点：爱尔兰，戈尔韦
设计师：菲利普·崔西
规模：101间客房
摄影师：图片由g酒店提供

Designed by Philip Treacy, the award-winning five-star g Hotel is one of Ireland's most luxurious spa hotels. Located in the heart of vibrant Galway, a city bursting with life, energy and vitality, and overlooking Lough Atalia, the hotel offers all that you would expect from an international five-star property and more.

The g's design director is renowned milliner and Galway native Philip Treacy and his special touch is evident everywhere, from the bespoke concierge wall to the artwork in every room. The emphasis is on luxury and glamour, underlining the commitment at the g to a truly unique hotel experience.

From the moment you step inside the g you know that you are in a uniquely special place, one which evokes not only old-fashioned Hollywood glamour but also the hotel's location near the romantic coastline of Galway. The dark and intimate reception area sets the tone, with black glass walls acting as a frame for a tank of Connemara-bred seahorses and bespoke concierge wall, designed by Philip Treacy. Covered in white Venetian plaster and reminiscent of a seashell, this is a hint of the surprises to come.

The public areas of the g on the ground floor extends to over 18,000 sq ft with three individually styled and themed lounges and a restaurant, all linked by a raspberry carpeted corridor. The impression is of walking through a particularly glamorous dolls house, with golden doorways framing the entrance to each room. From the exquisite Grand Salon to the distinctively "pink" Pink Salon and more masculine Blue Lounge, the g offers a stunning environment to meet and relax.

If the guest's introduction to the g hotel is through its wildly glamorous public rooms, the guestrooms are havens of intimacy and sensuality. Each guestroom has the distinctive Philip Treacy touch, evident in the detail as well as in the overall effect – which is one of stylish originality and supreme comfort. Philip's choices – from the wallpaper, lighting and artwork to the furniture and flooring – have been inspired by the landscapes and seashores of County Galway. The colours, textures and materials used reflect this heritage and tradition, but are also effortlessly contemporary. The g hotel offers a total of 101 guestrooms including 3 specialty suites – the Evangelista suite, listed in Elite magazine top 100 suites in the world, the Corner suite, the Atrium suite, 16 Luxury suites and 10 Junior suites. The remaining rooms are either of a superior or deluxe standard.

The g's distinctive approach to design is carried through to restaurant gigi's, with banquettes in purple velvet and jewel-coloured Andrew Martin chairs creating a dramatic and romantic atmosphere.

World leaders in spa design and product, ESPA has developed the total concept and design of this five-star ESPA at the g. ESPA at the g's design reflects a relaxed timelessness of contemporary elegance. In keeping with the overall design philosophy developed by Philip Treacy, nothing has been spared in its beautiful design and the superior quality of its finishes. Stone, wood, linen and glass are used in complementary new ways with subtle, warm colours. Mood enhancing lighting is used to calm and comfort, creating a sense of spiritual retreat.

这座由时尚设计师菲利普·崔西设计的五星级获奖酒店——g酒店是爱尔兰最奢华的水疗酒店之一。它坐落于充满活力的戈尔韦,这是一座生生不息,充满能量和生命力的喧嚣之城。酒店遥望阿塔利亚湖,提供给客人在一家五星级酒店所期待的一切甚至更多。

g酒店的设计总监是著名的女帽设计师——菲利普·崔西,同时他也是一位出生在戈尔韦的设计师,他的设计触及到酒店的方方面面,从定制的接待墙到每间客房中的艺术品。酒店的设计重点是奢华与魅力,设计师接受设计g酒店的初衷是想设计出真正独特的酒店体验。

从客人步入酒店的第一刻起,就会感受到这是一个独特的地方,它不但让人联想起好莱坞的旧时尚魅力,还让人真正的感受到此时正在戈尔韦浪漫的海岸边。昏暗幽谧的接待区奠定了整个酒店的氛围基调,海马水缸和定制的玻璃墙由黑色的玻璃墙作为基底,这些由设计师专门设计。墙上涂上了白色威尼斯灰泥,让人联想起贝壳,暗示着惊喜即将到来。

一层的公共区域面积达1800平方米,这里包括三个独立的时尚主题酒廊和一间餐厅,几个空间之间由铺着覆盆子图案地毯的走廊连接。让人印象深刻的是,在这家别具魅力的梦幻酒店穿行,每个空间的入口处都由金色的门廊作为标志。从精致的豪华沙龙到别具一格的粉色沙龙,再到充满阳刚之美的蓝色酒廊,g酒店为客人提供了最让人惊喜的会面和休闲环境。如果对客人来说最具魅力的是g酒店的公共空间,那么客房称得上是能带给客人亲切感和感官享受的天堂。每一间客房都体现出菲利普·崔西的独特风格,这既体现在细节上又体现在整体的效果上——这里是时尚原创性和极致舒适性的结合。菲利普对客房内的物品进行了精心的挑选,从墙纸、灯具、艺术品到家具地板,这些都是从戈尔韦郡的景观和海边风光获得的灵感。色彩、质地和材料用来呼应这里的传统,也同时无需费力的体现出现代感。g酒店提供总共101间客房,其中包括3间特色套房,被精英杂志列入世界100个顶级套房的伊万格利斯塔套房、转角套房、中庭套房,16间奢华套房和10间普通套房。其他的客房是高级客房和豪华标准客房。gigi餐厅的设计采用了最非凡的设计手法,紫色的天鹅绒长沙发搭配珠宝色的安德鲁·马丁设计的座椅,营造出充满戏剧性和浪漫气息的氛围。

设计顶级的水疗中心ESPA将他们全部的概念和设计理念带入g酒店的ESPA水疗中心。ESPA在g酒店的水疗中心设计体现出的是一种轻松氛围和永恒的优雅现代感。与菲利普·崔西的整体设计哲学保持一致,这里的设计同样体现出卓越的氛围和超高的质量。石料、木料、亚麻和玻璃在这里用现代的新颖方式被加上多变、温暖的色彩。灯光用来加强宁静、舒适的氛围,让这里成为精神上的休憩之所。

1. Thermal Suite in ESPA
2. The exterior of g hotel
3. Fashion cocktail bar
4. Pink Salon
5. Atrium Suite
6. g reception room
7. The bedroom in Deluxe King
8. Board room
9. Junior Suite

1. ESPA水疗中心热浴室
2. g酒店外部
3. 时尚的鸡尾酒吧
4. 粉色沙龙
5. 中庭套房
6. g接待宴会厅
7. 豪华国王套房的卧室
8. 董事会议室
9. 标准套房

# The Eternal Legend in May Fair Street

梅菲尔街上的永恒传奇

In the May Fair Hotel, an epitome of the whole London fashion world can be seen by the guests. This hotel, which was constructed by the end of 20s of last century, has gone through about a century's fashion time. Just the famous persons and fashion ICONs that are related to this hotel is numerous. At the opening of the hotel, King George V and Queen Mary participated in the ribbon cutting ceremony. And now Samantha Cameron, the British prime minister David Cameron's wife, is also tightly bound to this hotel. May Fair is the official hotel for the London fashion week that she had participated in. During the London week, all the brand shows and cocktail reception parties are all held in this hotel. The alliance between fashion and this fashion place was promoted through the love of fashion designers, fashion bloggers, and invited distinguished persons to May Fair. That's why fashion people love May Fair. The hotel interior which is renovated by designer Michael Attenborough should be a cardiotonic. The whole atmosphere around the hotel is fashionable, which makes it become one of the most attractable hotels for fashion people and one of the best places to take wonderful pictures for the magazines. Only the Fendi leather furniture in the hotel guestrooms promotes the hotel's fashion sense immediately. Furthermore, the fashion designers are often invited to design the show windows for the hotel. For example, during the London fashion week of this year, Annette Felder and Daniela Felder, who are the twins designers for the brand Felder Felder, were invited to decorate the hotel's show window with the theme of rock twins. When they talked about this creation, they said, "May Fair and Fendi is an eternal legend, whether in music, film or fashion."

在梅菲尔酒店，客人可以看到整个伦敦时尚界的缩影。这个在20世纪20年代末建成的酒店，经历了将近一个世纪的时尚风云。光是与酒店相关的名人和时尚ICON就数不胜数，在开业之初，就有国王乔治五世和玛丽皇后参与剪彩仪式，直到现任英国首相戴维·卡梅伦的妻子萨曼莎·卡梅伦也与酒店有着不解之缘，她参与合作的伦敦时装周的官方酒店正是梅菲尔酒店，伦敦周期间的所有品牌走秀和鸡尾酒招待会都将在酒店举行。时尚设计师，时尚博客主以及受邀的名流对于梅菲尔酒店喜爱促成了一次时尚与时尚之地的联姻。时尚界人士缘何对梅菲尔钟爱有加，由设计师迈克尔·阿滕伯勒重新翻修的酒店室内应该是一剂强心针，酒店随处的氛围都很时尚，这家最吸引时尚名人的酒店也成为时尚杂志拍摄的绝佳取景地。单是酒店客房内的FENDI皮革家具，就使酒店的时尚感立刻提升。此外酒店的橱窗经常邀请时尚设计师跨界设计，如在今年的伦敦时装周期间，酒店邀请Felder Felder品牌的双胞胎设计师安妮特·菲尔德和丹妮拉·菲尔德以摇滚双胞胎主题设计并布置了酒店的橱窗，当她们谈到这次创作时说道："梅菲尔酒店是一个永恒的传奇，无论是在音乐、电影和时尚方面。"

# The May Fair Hotel, London

伦敦梅菲尔酒店

**Completion/Latest renovation date:** 2006 /2012
**Location:** London, UK
**Designer:** Michael Attenborough
**Size:** 404 luxury rooms and suites
**Photographer:** The May Fair Hotel, London

**完成/翻新时间**：2006年/2012年
**项目地点**：英国，伦敦
**设计师**：迈克尔·阿滕伯勒
**规模**：404间客房和套房
**摄影师**：图片由伦敦梅菲尔酒店提供

The May Fair Hotel is an icon of style. Since the twenties, the hotel has been a London landmark and it continues to define luxury to a new generation of chic international travellers and Londoners alike. Following its 2006 revival, The May Fair has evolved into an edgy, fashionable and connected central London hotel.

The May Fair has an impressive list of "must see" rooms and inspiring experiences, making it one of London's finest hotels. Perhaps the most impressive aspect of the renovation project are the eight signature suites which offer the ultimate in glamour and style. The suites range from understated elegance to complete decadence, and offer colour palettes from a vibrant hot pink to muted naturals or should it read neutrals. Furniture from top designers, hand-chosen by the hotel design team, bespoke beds and individually designed hand-made carpets and furnishings, sourced internationally, add flair and class to some of the most stunning rooms in London.

The sweetest of suites are arguably the Azure, complete with a secret second entrance and private lift, the hot-pink Schiaparelli, one of the most breathtaking rooms in London, and the Penthouse, which comes equipped with its own bar and a roof terrace large enough to hold 70 people and offering stunning views over London. New to the collection is the Ebony Suite, which features the very latest TOTO bathroom technology – straight out of Japan. TOTO brings a slice of the future to The May Fair, with its illuminated bath, colour-changing, heat-sensitive basin controls and luxury, fully automated Washlet toilet, complete with remote control.

Other updates and additions include The Crystal Room, London's most spectacular event space with orient-inspired Shagreen style walls featuring the largest Baccarat chandelier in Europe.

There is also the newly added May Fair Theatre. The 201-seat private theatre/cinema is the largest of its kind in central London. The theatre skillfully mixes luxurious surroundings with cutting-edge technology.

The May Fair Bar, with a cocktail list to suit all crowds, is a must-visit venue. 2012 renovations include DJ residencies from internationally renowned musicians and gastronomic tailoring for London Fashion Week.

Above all, the hotel offers discretion of service that pre-empts requests – a true five-star service for one of Europe's finest properties.

梅菲尔酒店是一座极具时尚风格的偶像级酒店。自20世纪20年代开始，酒店就是伦敦的地标建筑并为新一代的国际客人和伦敦人定义了奢侈的概念。2006年重建至今，梅菲尔酒店已成为新锐、时尚、联通四方的伦敦最佳酒店。

梅菲尔酒店拥有令人印象深刻，给予新灵感体验的系列"必看"客房，使得它成为伦敦精品酒店。也许，改造项目中令人印象最深的是那8间魅力和风格达到极致的标志套房。从低调的优雅到极致的奢华，这些套房的色调也从充满活力的粉红到柔和的自然色或应称为冷色调。内部家具由顶尖设计师设计并由酒店设计团队亲选；定制床、单独设计的手工地毯和家饰均国际采购，这让这家伦敦最耀眼的酒店更具优秀资质。

梅菲尔酒店中最甜美的套房莫过天蓝色客房，内有暗门和私人电梯。最令人惊叹不已的套房则是艳粉色的夏帕瑞丽客房。阁楼客房配有酒吧和可容70人的屋顶阳台，在那里人们可饱览伦敦全貌。在这些精品

1. The lounge in Ebony Suite — 梅菲尔黑檀套房的客厅
2. Quince Restaurant — 昆西餐厅
3. The terrace of Ebony Suite — 黑檀套房露台
4. Cigar Room — 吸烟室
5. The sitting room in Duplex Suite — 双层公寓套房的起居室
6. The bedroom in Ebony Suite — 黑檀套房的卧室
7. The bedroom in Shiaparelli Suite — 夏派而丽套房的卧室
8. The lounge in Amarillo Suite — 阿美丽罗套房的休息室
9. The dining room in Ebony Suite — 黑檀套房的餐厅
10. The bedroom in Amarillo — 阿美丽罗的卧室

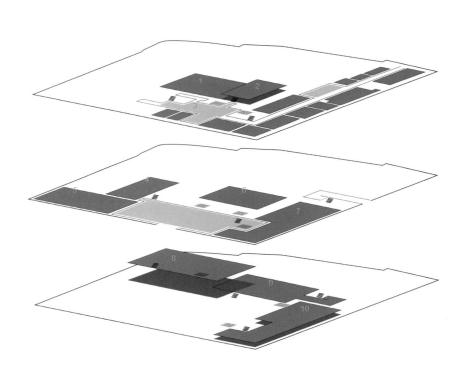

1. The Danziger Suite — 丹泽套房
2. The atrium — 中庭
3. Mezzanine Ground Floor — 中层楼 一层平面图
4. Amba Salon — 安巴沙龙
5. Amba Bar & Grill — 安巴酒吧及烤肉餐厅
6. Palm Beach Casino — 棕榈海滩赌场
7. May Fair Bar Lower Ground Floor — 梅菲尔酒吧 地下一层平面图
8. The May Fair Theatre — 梅菲尔剧院
9. The Crystal Room — 水晶客房
10. May Fair Spa — 梅菲尔水疗中心

客房中,黑檀套房最为新颖,其特点是直接从日本移入最新的TOTO卫浴技术,如光亮的浴盆、颜色更变、热敏性盆底控制以及可远程控制的豪华全自动卫洗丽卫生间。TOTO的这些卫浴技术带给梅菲尔酒店一丝未来气息。

其他的翻新和扩建工作包括水晶会议室,这是伦敦最壮观的活动空间,它有着受东方文化影响的粗糙的墙面,天花顶上装饰着全欧洲最大的百家乐水晶吊灯。酒店还新增加了梅菲尔剧场。这间有201个座位的私人剧场/电影院是伦敦中部地区同类剧院之最。剧院在技术上混合了奢华的氛围环境和最尖端的技术。

梅菲尔酒吧,提供适合大众口味的鸡尾酒单,这是一个到了梅菲尔一定要光顾的地方。2012年,酒吧进行了翻新,因此DJ区域得到了翻修,此外酒吧还为伦敦时尚周邀请到国际著名的音乐家和美食家。

总之,酒店提供周到的服务以满足顾客提前提出的要求,这确实是欧洲最好的五星级服务酒店之一。

# Hotels: Painting Rooms
酒店内的绘画空间

Artists create the most inspiring Castalia through painting and sculpture in spaces. Here displays the hotels possessing art elements. These hotels become the workshops of the artists where traditional patterns are broken and rules are redefined, including Gladstone Hotel designed by 37 artists; Casa do Conto, the fruit of history, culture and coincidence, provides the access to its true charm; Hotel Bloom, in which 287 paintings created by as many as 287 artists serve as the back walls of the rooms, shows the creative standard of European young artists.

艺术家们在空间内用绘画和雕塑等形式创意出最能激发灵感的源泉。呈现了几家带有相当艺术元素的酒店，这些酒店俨然成为了艺术家们的创作室，在这里模式被破解，规则也被重新定义。这其中包括直接由 37 位艺术家设计的格拉斯通酒店；通过历史、文化与巧合创作出的故事之家酒店让我们领略艺术创作的真正魅力；繁荣酒店更是邀请多达 287 位艺术家设计酒店客房的背景墙，287 幅作品也展示了欧洲青年艺术家的创作水准。

# A Micro World of European Arts

欧洲艺术的微型世界

ELIA — European League of Institutes of the Arts — is an organisation which represents almost 350 institutes of the arts and covers 47 countries. It was founded in Barcelona, Spain in October 7, 2000. This organisation covers almost all kinds of art forms including fine art, design, performance, music and dance, media art and architecture. It is in charge of such relevant affairs as using membership to find young trendy artists, and setting up a selection committee to establish the special standards for choosing artists. The establishment of ELIA defines again the importance of art in people's life. It makes people turn their attention to the state-of-the-art of the higher education of art. It also becomes an open platform to connect the higher education of art to commercial practice. In the principle of jointly driving the art education of European countries, ELIA continually holds different kinds of activities and exhibitions which can help young artists majoring in arts find the right road of development in their career.

Hotel BLOOM! is an example in which ELIA integrates arts with commerce. It gathers 287 artists who are specialised in different fields. All the guestrooms use the same colour of white and the same decoration method. The only difference is that they have different sizes and different frescos. The young artists from different countries use their own frescos to express their own art thoughts. They come from different countries, and express their roots and experiences in their works. The art soul of the hotel is weaved by these frescos which constitute a micro world of European arts.

ELIA，欧洲艺术学院联盟———一个代表将近350个艺术类高等院校、覆盖47个国家的艺术组织，2000年10月7日在西班牙的巴塞罗那成立，组织涵盖几乎各种艺术形式，其中包括美术、设计、表演、音乐舞蹈、媒体艺术和建筑。组织负责的艺术相关事务包括：使用会员制发掘欧洲新潮的年轻艺术家，建立选拔委员会以建立选择艺术家的特殊标准。ELIA的成立再次定义了艺术在人们生活中的重要性，让人们开始关注艺术高等教育的发展现状，也成为将艺术高等教育与商业实用联系到一起的一个开阔的平台。联盟将共同推进欧盟国家的艺术教育为原则，不断的举行各类活动和展览，促进艺术专业的年轻艺术家在事业上尽快找到正确的发展之路。

繁盛酒店是ELIA将艺术与商业接轨的一个实例，它为繁盛酒店提供了287位专攻于各个领域的艺术家为酒店进行创作，所有的客房都使用相同的白色调和相同的装饰方法。不同之处只是有着不同的大小和不同的壁画。来自不同国家的青年艺术家用各自的壁画表达着自己的艺术思想，他们来自不同的国家，也从他们的作品中表达出他们的根源、他们的经历，酒店的艺术灵魂由这些壁画交织而成，组成了一个欧洲艺术的微型世界。

# Hotel BLOOM!

繁盛酒店

**Completion/Latest renovation date:** 2007/2012
**Location:** Brussels, Belgium
**Size:** 305 rooms
**Designer:** DOOS / architect: Bronwynn Welch
**Photographer:** Luk van der Plaetse

完成/翻新时间：2007年/2012年
项目地点：比利时，布鲁塞尔
规模：305间客房
设计师：DOOS设计事务所，布伦韦恩·韦尔奇
摄影师：卢克·范·德·普利茨

Hotel BLOOM! is a truly modern hotel in central Brussels, Belgium, where every room is decorated differently with a unique fresco made by a young European artist. Choose the happening place where guests can relax and totally be themselves. Whether people are here for pleasure or business, in the lobby, in the rooms, in SmoodS living room, kitchen and musict, even in one of the 13 meeting rooms… guests can get an experience that is inspirational.

In 2007, Hotel BLOOM! renewed the hotel and invited 287 creative young artists from different European countries to paint the rooms of the hotel and make it something unique for the guests. Now guests have the luxury to choose their own favourite room with a hand-painted fresco on the wall by their own interpretation of the word "bloom".

The hotel is located near the botanical gardens. Lots of flowers, plants, trees and herbs blossom the whole year. "To bloom" means "to open", "to grow or to flourish", which totally echoes the building of the hotel.

Overall, at BLOOM! there are no stuck-up rules. Forget dress codes, room service or bell boys. It attracts people that have an open mind. Because it does things differently!

Indeed, at Hotel BLOOM! guests stay away from the traditional and conservative hotel experience. Upon entering, the difference is immediately noticeable: this is very different. Because Hotel BLOOM! is modern, fresh and contemporary.

The lobby lounge is spacious and gives the impression that guests can breath freely after leaving the hurlyburly of the town and the rue Royale. Guests get a feeling of relief, freedom and refreshment.

Hotel BLOOM! has 12 stylish, bright meeting spaces. Wired to meet all technology needs, along with multi-coloured squares of blackboards along the walls that look more Mondrian than mundane. Breaks for coffee and otherwise are relaxing, recharging with Wii to play and a great lounge, for, well, just lounging.

SmoodS, the restaurant and bar of Hotel BLOOM! in Brussels, is a cosy, stylish and trendy living room. This is an original concept, which creates a familiar feeling at the same time. The setting includes 7 different atmospheric islands which have been created around a central bar: Passion, Bazaar, Safari, Bling-Bling, Aqua, Spring and Flower Power.

At SmoodS choose the atmospheric island that suits the mood and where guests can stay to eat and drink something. Or where guests can sit back for the perfect relaxing moment and listen to the resident DJ's music. SmoodS has an open kitchen. Especially from the atmospheric islands Bling-Bling, Flower Power and Spring guests have a good view of what happens in the modern, black-tiled kitchen.

繁盛酒店是一家真正的现代酒店,它坐落在比利时布鲁塞尔的市中心,这里的每一间客房都由一位年轻的欧洲艺术家装饰,每一位艺术家都为一间客房创作壁画。在这里客人可以放松去做他们真正的自己。无论是在这里娱乐还是谈生意,无论是在大堂、客房、餐厅、厨房,甚至是在13间会议室的任意一间,客人们在这里感受到的是一次充满灵感的体验。

2007年,繁盛酒店焕然一新,酒店邀请287位来自欧洲各国的具有创意的年轻艺术家,为酒店的每间客房绘制油彩,让客人感受到独一无二的艺术气息。如今,客人有机会能够选择带有他们最喜爱的油彩壁画的客房,这就是他们在酒店里用到"繁盛"一词的原因。

酒店离当地的植物园很近。全年有各式的花、植物、树木和草本植物盛开。"To Bloom"意思是"开启","繁荣的生长",这正是酒店建筑的回应。

总之,在繁盛酒店,客人不会受到规则的束缚。忘记那些着装规则、客房服务和门童。酒店用开放的理念吸引客人,一切源自与众不同。

在繁盛酒店,客人可以远离传统、保守的酒店体验。一踏入这里,现代、清新、时尚的繁盛酒店立即就会让人感受到与众不同。

大堂休息区非常宽敞,且令人印象深刻。当客人急匆匆的从城区和皇家大道游玩返回,可以在这里放松休息、享受自由、振作精神。

繁盛酒店有12个时尚、明亮的会议空间。空间内布置了充足的电路满足各类电子设备的需要,墙上挂着彩色的写字板,很有些莫德里安绘画风格的意思。这里提供咖啡和其他饮品让客人放松充电,酒店还提供娱乐服务,在漂亮的会议休息区, 客人在这里可以选择玩任天堂游戏,也可以仅是休息。

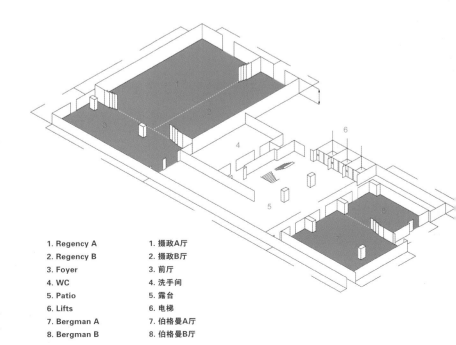

| | |
|---|---|
| 1. Regency A | 1. 摄政A厅 |
| 2. Regency B | 2. 摄政B厅 |
| 3. Foyer | 3. 前厅 |
| 4. WC | 4. 洗手间 |
| 5. Patio | 5. 露台 |
| 6. Lifts | 6. 电梯 |
| 7. Bergman A | 7. 伯格曼A厅 |
| 8. Bergman B | 8. 伯格曼B厅 |

心情S餐厅和酒吧是一间惬意、时尚的就餐空间。这里是一个原创设计的空间，营造出轻松的氛围。这里有7种不同氛围的休息岛围绕着中心酒吧布置，主题分别是：激情、集市、旅行、亮晶晶、水下世界和花朵力量。

在心情S餐厅，客人选择适合他们心情的休息岛就餐或者喝饮料，或者坐着放松休息，享受DJ播放的音乐。心情S餐厅内有一间开放式厨房。特别是坐在亮晶晶、花朵力量和春天这些休息岛，客人可以清晰的看见这家现代的、贴满黑色瓷砖的厨房里正发生着什么。

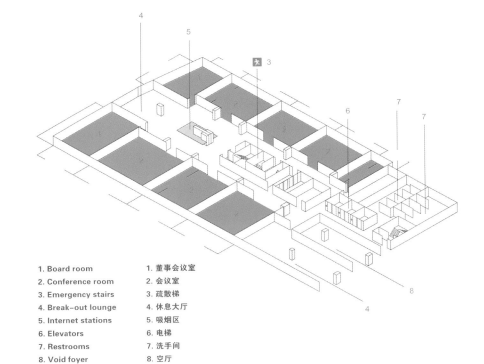

1. Private lounge in SmoodS
2. The reception
3. A corner in the corridor
4. The foyer in meeting space with screen
5. The corridor to meeting rooms
6-17. The guestrooms with fresco created by young artists in ELIA

1. 思慕斯餐厅的私人休息区
2. 接待区
3. 走廊一角
4. 会议空间有屏幕的前厅
5. 通向会议室的走廊
6-17. 欧洲艺术学院联盟的青年艺术家创作的客房壁画

| | |
|---|---|
| 1. Board room | 1. 董事会议室 |
| 2. Conference room | 2. 会议室 |
| 3. Emergency stairs | 3. 疏散梯 |
| 4. Break-out lounge | 4. 休息大厅 |
| 5. Internet stations | 5. 吸烟区 |
| 6. Elevators | 6. 电梯 |
| 7. Restrooms | 7. 洗手间 |
| 8. Void foyer | 8. 空厅 |

# Art Graffiti for Toronto

多伦多的艺术风情涂鸦

Gladstone Hotel is internationally recognised as one the best art hotels in Canada. Here, the juxtaposition of the new and the old, history landmark and modern art, produces a shocked contrast effect, which makes Gladstone Hotel become the mark of Toronto culture and the bridgehead for international guests to visit Toronto. Gladstone Hotel is a commercial project based on values. The hotel devotes itself to provide guests with an immersive and authentic experience of Toronto landscape. Christina Zeidler is the owner and CEO of the hotel, who is in the meantime an artist and film maker. Since 2003, she had adopted the form of community participation to renovate Gladstone Hotel and endowed it with new vigour. She had invited 37 local artists and designers to design and finally realised their guestroom design. On the premise that the usage function is guaranteed, each guestroom's design is unique, which is custom-made according to the designers' style and imagination. Each guestroom is a little world with its own tune and story which can cater to different guests' likings. D'arch St. Pierre, the designer of room 308, the Queen Suite in Queen Street, is a dress shirt designer. His inspiration came from the long history of Gladstone Hotel's building culture and the vigorous art design district of west Queen Street. Back against the red brick wall and faced with the colourful and abundant scrawls, guests can not only feel fresh and stimulating, but also have a feeling of intimacy. Room 405, the twilight reception room, is designed by sculptor Corwyn Lund collaborated with Simone Moir (a performance, screenage and device artist). Misted neon light, shiny reflective smooth textile, tinfoil ceilings embossed with primitive decorative patterns, exquisite blue chandeliers, indistinct plant patterns on the wall, and the soft light of white, gold and red irreflexive on the walls and ceilings, combined with furniture of 40s style, make the room brim with a fascinating mystique of the new black film.

Other rooms are the same as art experience space full of imagination, or amazing graffiti created by reckless inspiration. Just as D'arcy St.Pierre said: "We want to move the Toronto alley that is painted with scrawls to room, which can fuse the history sense of Victoria Age and modern city culture together."

格拉德斯通酒店是国际公认的加拿大最好的艺术酒店。在这里，新与旧相融，历史地标与现代艺术并列，产生震撼的反差效果，这使得格拉德斯通酒店成为多伦多地方文化的标志以及国际旅客游览多伦多的桥头堡。格拉德斯通酒店是一个具有艺术价值观的商业项目。酒店致力于为旅客们提供沉浸式的、地道的多伦多街头艺术风情体验。克里斯蒂娜·蔡德勒是酒店的所有人兼总裁，同时她也是一位艺术家和电影制作人。自2003年，她采取社区共同参与的形式翻新格拉德斯通酒店并赋予它新的活力。她邀请37位本地艺术家和设计师设计并最终实现他们的客房设计。在保证使用功能的前提下，每间客房的设计都是独一无二的，设计师根据自己的风格、想象力，度身定制每间客房。每间客房都是一个小天地，有自己的调调，自己的故事，可以满足不同客人的不同喜好。308房间——皇后街皇后套房的设计师达西·圣皮埃尔是位礼服衬衫设计师。他的灵感来自于格拉德斯通酒店悠久的建筑文化历史以及皇后街以西充满活力的艺术设计区。旅客背靠红砖墙，面对色彩鲜艳，内容丰富的涂鸦，既感到新鲜刺激，又能体会到一种亲切感。405房间暮光会客室是雕塑家戈恩·隆德和西蒙尼·姆瓦（表演、影像以及装置艺术家）共同的杰作。迷离的霓虹灯光、盈盈反光的光滑织物、凸印古朴花纹的锡箔天花顶板、精致的蓝色垂穗吊灯、墙壁上朦胧的植物图案以及墙壁、天花板漫反射的柔和的白色、金色和红色的光，再配以20世纪40年代风格的家具，整个房间洋溢着新黑色电影迷人的神秘气氛。

接下来的其他房间同样有如一个个充满想象力的艺术实验空间，一幅幅肆意创作的涂鸦作品，就如达西圣·皮埃尔所说："我们想把绘满涂鸦的多伦多小巷搬进房间，让维多利亚时代的历史感与现代城市文化融为一体。"

# Gladstone Hotel

格拉德斯通酒店

**Completion/Latest renovation date:** 2005
**Location:** Toronto, Canada
**Designer:** Christina Zeidler and 37 local artists
**Size:** 37 Artist Design Hotel Rooms
**Photographer:** Courtesy of Gladstone Hotel

完成／翻新时间: 2005年
项目地点: 加拿大,多伦多
设计师: 克里斯蒂娜·蔡德勒及37位本土艺术家
规模: 37间客房
摄影师: 由格拉德斯通酒店提供

The Gladstone Hotel's 37 artist-designed guest rooms are each truly distinct works of art, designed by a different artist in the local community. That means that each room is in a different style, catering to the tastes of any traveller.

The Gladstone Hotel is a perfect gateway to the Toronto arts scene with artist-inspired rooms, creative exhibitions, events and food that all reflect the diversity of talent in Toronto.

The artist-designed room project was created by the Gladstone Hotel's owner and President Christina Zeidler. Zeidler is an artist and filmmaker, who has a passion for sustainability and social activism. When she created the project she put out a call for submissions and allowed artists, designers, architects and craftspeople from the local arts community to submit design proposals for each of the 37 rooms. A jury comprised of local artists and designers selected the final 37 proposals and the participating artists finished each room.

Each room is based on the artist's vision while conforming to the needs of the hotel guest. Each of the 37 rooms reflects the diversity of talent in the city of Toronto. The rooms give a taste of the authentic flavour of Toronto to guests who are new to the city and resonate with guests who know the city well.

Each of the incredible artist-designed guest rooms is truly distinct, presents guests with a unique experience and is designed by a different artist. The word "artist" is used broadly here to encompass a number of disciplines and includes visual artists, interior designers, architects and material-based artists. The rooms are based on the artist's vision while conforming to the needs of the hotel guest.

Not only are the 37 hotel rooms artist-designed, but the entire building is dedicated to art with a full-time Director of Exhibitions, over 90 exhibitions a year, a permanent collection and three exhibition spaces. The Gladstone Hotel self-produces several major group exhibitions a year, giving artists and designers the opportunity to have their work presented to a broad and engaged audience in an unconventional setting. The most popular of these exhibitions is Come Up To My Room, which hosts over 30 artists and 3000 people in 4 days. Additionally the Gladstone Hotel offers exhibition rental space where artists and art collectives can put on their own self-produced shows.

The Gladstone also curates and self-produces many music events, giving the opportunity to local musicians and sound artists to present their work to a large audience.

Finally, the Gladstone donates venue space to artists, artist collectives and community groups in order to support the local arts community. All of these initiatives are part of their mandate to support, facilitate and incubate their local arts community in order to make a stronger community.

1. Snapshot designed by Christina Zeidler
2. Teen Queen designed by Cecilia Berkovic
3. Felt Room designed by Kathryn Walter
4. Trading post by Matthew Agostinis and Joel Harrison
5. Canadiana Room designed by The Big Stuff and Jenny Francis
6. Tower Suite designed by Christina Zeidler and Jane Zeidler
7. Offset designed by Heather Dubbeldam and Tania Ursomarzo
8. Queen Suite on Queen Street designed by Maison St. Pierre
9. Flight316.CA designed by Adam Berkowitz and George Simionpoulos
10. Urban Voyageur designed by Koma Designs and Paul Fortin
11. Melody Bar
12. Gladstone Café
13. The tables in Melody Bar
14. The details in Frank designed by Paul Campbell

1. 快照客房由克里斯蒂娜·蔡德勒设计
2. 青春皇后客房由瑟西莉亚·贝尔科维奇设计
3. 感觉客房由凯瑟琳·瓦特设计
4. 商栈客房由马修 阿格斯汀尼斯和乔尔 哈里森设计
5. 加拿大人客房由大团队设计事务所和珍妮·弗朗西斯设计
6. 塔楼套房由克里斯蒂娜·蔡德勒和珍妮·蔡德勒设计
7. 抵消客房由海瑟·杜伯丹姆和谭妮亚·乌索玛佐设计
8. 皇后街上的皇后套房由圣皮维尔之家设计事务所设计
9. 加航316航班客房由亚当·伯克维兹和乔治·西蒙普鲁斯设计
10. 都市航行客房由科马设计事务所和保罗·佛丁设计
11. 旋律酒吧
12. 格莱斯通咖啡厅
13. 旋律酒吧餐桌
14. 由保罗坎贝尔设计的弗兰克客房细节

由于每间客房都由本土的不同设计师设计，因此格拉德斯通酒店的37间艺术设计型客房是货真价实的艺术珍品。每间客房风格迥异，能迎合各类游客的品位。

坐拥富有艺术灵感的客房、拥有足以展现多伦多济济人才的创造力的展品和美食，格拉德斯通酒店为游客铺设了一条通往多伦多艺术全景的完美之路。

该艺术设计客房项目是由酒店所有人兼总裁克里斯蒂娜·蔡德勒建立。克里斯蒂娜·蔡德勒是一位艺术家兼电影制作人，长期热衷社会活动。当开始这一项目时，她对外征求意见并允许当地艺术圈内的艺术家、设计师、建筑师和工匠们为每间客房提交设计方案。之后由当地艺术家和设计师组成的评委会从众多提交方案中选出37个最终方案，并由参与设计的艺术家独立完成。

在满足酒店客人需求的同时，每间客房都渗透出各自的艺术视角，折射出多伦多城的人才济济。这里的客房为初到多伦多的游客提供了纯正的多伦多艺术风味，而对于熟悉这一城市的人，则唤起了他们内心的共鸣。

每间艺术设计型客房巧夺天工、独具匠心；不同艺术家为客人提供了独特体验。此处的"艺术家"一词指代丰富，包括视觉艺术家、室内设计师、建筑师和导向艺术家。

除37间客房的艺术雕琢外，酒店的整体建筑也是艺术的化身。酒店设有全职馆长，负责永久展品的保养和3个展区、年逾90场展览的工作。格拉德斯通酒店每年自办几大巡展。借以酒店非传统的场地布局，为艺术家和设计师提供了向广大热心观众提供作品展示的良机。其中尤以"来我的房间"这一展览最受欢迎，4天中共有30余名设计师参与展出并吸引了3000名游客前来参观。此外，格拉德斯通酒店还出租展览空间，允许艺术家和艺术团体展出他们的原创作品。

格拉德斯通酒店也自创很多音乐活动，为当地音乐家和声乐艺术家提供了向大众展示其作品的良机。此外，格拉德斯通酒店支持当地的艺术，为艺术家、艺术团体和社会团体提供场地，这些积极举措是我们支持、营造、培育并进一步强化当地艺术的一个组成部分。

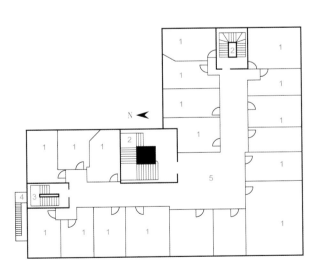

1. Room
2. Elevator & Stairs
3. Stairs
4. Fire escape
5. Corridor

1. 客房
2. 电梯及楼梯
3. 楼梯
4. 防火通道
5. 走廊

# The Sculpture Graven by Scar

伤疤雕刻的艺术品

Casa do Conto, arts and residence is a work of art in itself. It quietly lies there as a story and a sculpture of soul. From the view of its name, this hotel combines arts with residence. When we stand in it, we can hear its existence and breath. It is a typical example that puts arts into practice. The hotel is situated in Porto Portugal which is a city of art and famous for its architectures. There are well-preserved Baroque architectures in this harbour which hadn't been through any natural disasters or wars. And on the other side of this city which is influenced by the design renaissance, a lot of architectures with artistic charm make up another landscape painting. And Casa do Conto, arts and residence was born here like a miracle.

The designer had to make a new design proposal for this hotel because an accidental fire. And the artistic charm of the hotel was achieved by this scar. Six artists' stories were carved in the interior of this architecture which is like a fossil. The designer was determined to integrate the image of the fire and ashes into this new design. However, he didn't intend to make it too sorrowful and solemn but convey a message of perseverance, strength, and rebirth. It provided a significance of deeper levels for the guests' lodging experience while engraved this destructive disaster on their mind. "It's just like a phoenix that can produce miracle from the ashes. The story about this architecture makes it more special and endows us with another kind of inspiration," a designer from Pedra Liquida described their design of this hotel. Destroy, hope, and then dream come true. This is just the real charm of art.

故事酒店本身就是一个艺术品。它静静的躺在那里，是一段故事，一座灵魂的雕塑。从名字来看酒店将艺术和住宅结合起来，当我们站在其中，聆听它的存在与呼吸也正是一个将艺术予以实用的典型例子。酒店坐落在葡萄牙的波尔图，这是一个本身就以建筑闻名的艺术之都，这座没有遭遇过天灾和战争的港湾，有保存完好的巴洛克式建筑，而城市的另一面近些年来受到设计复兴风潮的影响，许多充满艺术魅力的建筑也组成了另一幅风景画，故事酒店在这里也奇迹般的诞生了。

酒店因一道伤疤成就了它的艺术魅力，一次意外的火灾让设计师规划出新的设计方案。化石般的建筑在内雕刻上6为艺术家的故事。设计师决心在新设计中融入火灾以及灰烬的意象，但并不想显得过于悲伤深重，而是传达坚强、力量、重生的寓意，在铭记这次毁灭性的灾难之余，为客人的住宿体验增添更深层面的意义。"就像凤凰涅磐一样，从灰烬中会诞生奇迹，围绕这栋建筑的故事让它显得更加特别，也赋予了我们别样的灵感。"液体岩石工作室的设计师这样形容他们对酒店的设计。毁灭，希望，梦想成真，这也正是艺术的真正魅力所在。

# Casa do Conto, arts&residence

故事酒店，艺术与住宅

**Latest renovation date:** 2011
**Location:** Porto, Portugal
**Designer:** Pedra Líquida
**Size:** Six guestrooms
**Photographer:** Fernando Guerra FG+SG

翻新时间：2011年
项目地点：葡萄牙，波尔图
设计师：液体岩石工作室
规模：6间客房
摄影师：费尔南多·格拉FG+SG摄影事务所

Three years ago, Pedra Líquida (Liquid Stone) was commissioned to create a new hotel design concept – Casa do Conto, arts&residence – giving life to a beautiful XIX Century Oporto House, through a chirurgical restoration process. Unfortunately, on March 2009, few days before the hotel opening, the building suffered a terrible fire.

The architects and concept creators decided that they had to rebuild it, better than before. In fact, recreating its remaining structure was an opportunity to make this new hotel, and the memory of that special house, reborn from the ashes, like the Phoenix.

In this sense, the new project evokes, through an abstract approach, the old house adornment and its wall textures by using traditional surfaces – crossed wood patterns, corrugated steel plates and curved plywood panels – as a "mould" for the new concrete walls: at the central staircase; at the back facade; at the cubic bathrooms inside every suite; at the oval-shaped central skylight, a typical Oporto typology. As a result the designers get a kind of "fossilised architecture" where those modern "skins" rephrase the pre-existing ones.

All the concrete ceilings are (re)decorated by carved texts, in bass-relief, where you can read different narratives about the concept of "house" and of that house in particular. Created by different authors related to Oporto and its architecture, those six tails were spatially layered by Pedra Liquida, and graphically imagined by R2 Designers.

Casa do Conto embodies a unique story of life, which is, in fact, the history of the city itself. The hotel highlights Oporto's domestic architecture, solemn and vertical, carved by the scars of a time that mediates its decline and rebirth, the memory of the past and the desire for the future, the granite stone of the old facades and the new concrete of its inner core.

三年前，液体岩石设计工作室受到委托为一家新酒店——故事酒店，艺术与住宅进行概念设计，通过对其进行外科手术式的翻修，赋予这座19世纪的建筑美丽的新生。不幸的是，2009年3月，就在酒店开业的前几天，建筑发生了可怕的火灾。

建筑师和设计创意者们觉得重新建造这座酒店，而且要建的更好。事实上，在遗留建筑的基础之上再建酒店是让酒店重获新生的一次机会，这座特殊房子的记忆将像凤凰一样，在灰烬中重生。

在此基础之上，新的酒店以一种抽象的方法被唤醒了，旧建筑的装饰部分和墙体质地使用传统的方法建造——木纹路、花纹钢板和曲面胶合板作为一种新的模式运用在新的混凝土墙内，包括中央楼梯、外立面的背面、每间套房立方体浴室的内部、中央椭圆形天窗，都运用了这种波尔图的建筑方法。最后，建造出一种"石化的建筑"，这些现代的墙面"皮肤"令人回想起之前的酒店。

所有的水泥天花板被刻上了浮雕装饰文字，客人可以通过这些叙述文字读到房子的理念和特殊之处。文字由不同的作者创作，内容与波尔图和它的建筑相关，由六位建筑师设计了不同的故事，图文设计由R2设计工作室负责。

故事之家酒店体现的是生命中的一个特别的故事，事实上也是这个城市自己的历史。酒店强调了波尔图的民用建筑，它庄严，成垂直状，它拥有旧的花岗岩外立面和新的水泥内部核心，时间在它身上刻出刀疤也暗示着它的衰退与重生，过去的记忆和对未来的渴求。

1. The central stairs
2. The bedroom in Suite Residence NG
3. The kitchenette in Suite Residence AT
4. The bedroom in Suite Residence AD
5–7. The details of stairs and corridors
7. The bed in Suite Residence AT
8. The close-up for ceiling
9. The open ceiling in the corridor
10. The lounge
11. The reception
12. The restaurant
13. The details in bathroom

1. 中央楼梯
2. NG住宅套房的卧室
3. AT住宅套房的小厨房
4. AD住宅套房的卧室
5–7. 楼梯和走廊的细节
8. AT住宅套房的卧床
9. 天花板特写
10. 走廊里的开放式天花板
11. 休闲区
12. 接待区
13. 餐厅
14. 浴室细节

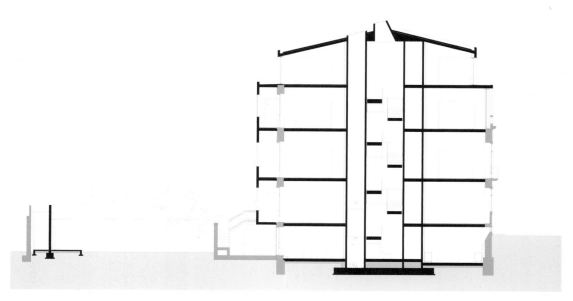

1. Bathroom
2. Bedroom
3. Stairs
4. Courtyard
5. Kitchen
6. Dining room
7. Restaurant
8. Toilet
9. Bar

1. 浴室
2. 卧室
3. 楼梯
4. 庭院
5. 厨房
6. 餐区
7. 餐厅
8. 洗手间
9. 酒吧

# Hotels as Art Works
酒店艺术品

An art piece is a small part in hotel design. However, these hotels improve the level by using art pieces and artworks that become the protagonist for creating atmosphere. Here introduces some hotels featuring artworks to enhance interior design atmosphere, including the Olsen Hotel where the famous Australian artist John Olsen's paintings are hung all around; the Crosby Street Hotel in New York is a hub where guests can amazingly find many British artworks; the citizenM Hotel, the mobile soul of art at the London bank is filled with artworks; Byblos Art Hotel Villa Amistà is infused with Baroque and its manager converted the hotel into an exhibition room of personal collection where art is perfectly blended with Baroque design.

艺术品在一般的酒店只是装饰部分的一个小环节，而这里介绍的几家酒店将艺术品的运用提高了一个层次，艺术品成为了制造氛围的主角。这里介绍了一些通过艺术品来加强室内设计氛围的酒店，这其中包括挂满了澳大利亚著名艺术家约翰奥尔森绘画作品的奥尔森酒店；在纽约的克罗斯比街酒店，客人会惊喜的发现这是一个英式艺术品的集散地；在艺术品无处不在的伦敦河岸区居民 M 酒店，艺术在这里成为了移动的灵魂；充满巴洛克风情的比布鲁斯艺术别墅酒店，这里的管理者将这种酒店变成了私人收藏艺术品的展示馆，让各类艺术与巴洛克创造出奇迹的交融。

# Salute John Olsen, Salute Australian Art

向约翰·奥尔森致敬,向澳洲艺术致敬

From the name of The Olsen Hotel, it is easy to know the design inspiration of the hotel. It is just the art and artist John Olsen who can represent the highest level on painting. John Olson is one of the most outstanding living artists in Australia. His works are collected in the Australian Art Museum and many other and local art museums. In recent years, his works are getting more and more popular and are the delight of personal and enterprise collectors at home and abroad. John Olson uses abstract method to describe the fascinating landscapes of the Australian mountains and rivers, desert and harbours in his paintings. The creative works are colourful, many of which set sand yellow as their main tone; and the technique of painting is original, sincere, emotive, and attractive. When appreciating the paintings, people feel like that they place themselves in Australia's magnificent nature. Olson Art Hotel has a good location that is at the end of the church, and close to the CDB of Melbourne, where there is dense atmosphere of art, fashion, and culture. The design of Olaon hotel is centred on Olson's art works, which makes it fashionable, lightsome, and full of anima. There is a huge work painted by Olson on the outside surface of the hotel. The colours used in the interior design are simple and concise, which are either modern silver grey or the representative sand yellow. The furniture and ornaments are conformably of the modern simple styles without complicated ornaments. The whole space of the hotel is like a modern art museum. The protagonist is Olson's paintings which are elaborately arranged in the full position of each space. From the outside to the interior of the hotel, guests experience the art atmosphere all the time. It's like a wild Australian travel led by John Olson.

从酒店的名字——奥尔森酒店,就能领悟到酒店设计时的灵感源泉,那就是艺术和象征着澳大利亚最高绘画水准的艺术家约翰·奥尔森。约翰·奥尔森——澳大利亚在世的最杰出的艺术家之一,其作品被澳大利亚国家美术馆、州立及地区美术馆收藏。近年来其作品也越来越受到国内外个人和企业收藏者所喜爱。约翰·奥尔森的画作用抽象的手法描绘澳大利亚山川河流、沙漠海港的迷人景象。创作用色多姿多彩,很多作品以沙黄色作为基调;笔法原始、挚诚,另人感动,诱人联想。人们观赏画作之时仿若穿过一片空灵置身于澳大利亚壮丽的自然之中。奥尔森酒店地理位置优越,位于教堂街尽头,临近墨尔本CDB,艺术、时尚、文化氛围浓厚。奥尔森酒店的设计围绕约翰·奥尔森的艺术作品展开,时尚轻盈,灵气十足,酒店外表皮上有一幅巨大的约翰·奥尔森画作。室内设计用色单纯简练,或为现代的银灰色或为奥尔森代表性的沙黄色。家具、装饰品一致的选用现代简约的风格,没有繁复的装饰。整个酒店空间好似一座现代美术馆,而主角就是奥尔森的画作。他们被精心地布置到各空间的最醒目位置。旅客由外部进入酒店内部无时无刻不在接受着艺术的气息,仿若踏上约翰·奥尔森引导的狂野澳洲之旅。

# The Olsen Hotel

奥尔森酒店

**Completion date:** 2010
**Location:** Melbourne, Australia
**Designer:** Chris Hayton
**Size:** 2800 m²
**Photographer:** Art Series Hotels

完成时间：2010年
项目地点：澳大利亚，墨尔本
设计师：克里斯·海顿
规模：2800平方米
摄影师：艺术系列酒店

Architect Chris Hayton, Principal at Rothelowman said John Olsen's art provided a "solid starting point" for "conceptual thinking" when designing The Olsen.

He explained that their design of an opaque glass, lightweight delicate skin, or canvas, draped over the hotel facade, served a dual purpose. "The lightweight, white curtain wall alludes to a tent-like structure and the idea of short-term accommodation that this conveys. Its uniqueness also assisted us in creating a boutique, destination hotel," Hayton said. "In addition, it evokes the idea of an artist's canvas and we feel all these references complement each other to create a major hotel at a renowned South Yarra location," Hayton said.

Hayton explained that their design included four different Olsen artworks to appear on the windows that would stand alone as complete works when seen from within the interior of the hotel but that from a distance, the façade would appear as one large, stand-alone artwork.

"We chose artworks for the windows that used a lot of linework rather than paintings that could have obscured the view," Hayton said. "We did a lot of research into Olsen's work to discover his inspiration and found a consistent fineness and elegance in his work that we carried over into the building design. Hotels are normally quite heavy designs but The Olsen is a celebration of fine lines, delicacy and a lightness of expression evident in Olsen's artwork," Hayton added.

Hayton said that the architecture constantly drew upon Olsen's work without being too literal. "The architecture had to come from our soul as well, so striking the right balance was paramount."

The Olsen Penthouse offers 160 square metres of tranquil and contemporary living space. With two private bedrooms, king sized Art Series signature beds, private ensuites, walk in robes, gas fireplace, three 42' HD flat screen TVs, full gourmet kitchen and dining table for 10, it is your own private sanctuary. Take pleasure in spectacular sweeping Melbourne CBD skyline views or soak your worries away in the eight-seater balcony spa while looking out over Chapel Street's thriving art, fashion and culture scene.

罗瑟劳曼设计工作室的负责人、建筑师克里斯·海顿声称，在设计奥尔森酒店时，他的设计理念思维深植于约翰·奥尔森的艺术。

他解释说，酒店外立面单反玻璃和轻质细腻的涂层或画布设计有着双重作用。"轻质的白色幕墙使人联想到竹帐，意识到此处为临时居所。其独到之处又助我们设计了精品酒店"，海顿补充说："此外，轻质的白色幕墙又唤起了我们对画布的联想。我们认为在著名的南亚拉地区建造大型酒店会使这些借鉴和参考相得益彰。"

海顿解释道，他们的设计包括将4件不同的奥尔森作品布置在窗口，当客人们站在酒店外观赏室内，所有的艺术作品集中在一起非常惹眼，而从远处观看，外立面呈现出是一个巨大而独立的艺术品。

"我们为橱窗选择的艺术品是一些素面作品而不是那些会吸引人注意力的油画"，海顿说道，"我们对奥尔森的作品进行了大量的研究，以便发现他的创作灵感，把其融入到建筑的设计之中，达到与他的作品一样的美好和优雅。酒店设计是一项相当难的设计工作，但是奥尔森是运用线条表现精细的专家，在他的作品中可以见证对轻盈感的表达"。海顿说，这座酒店建筑不断的吸取奥尔森作品的精华，且避免了形式上的照搬。"建筑也应该来自于我的灵魂，重要的是那些惊人精确的平衡性。"

奥尔森阁楼有160平方米宁静、现代的居住空间。两间私人卧室，超大号的艺术系列标准卧床，私人浴室，步入式衣橱，燃气壁炉，三个42英寸高清平板电视，装满美食的厨房和可供10人就餐的餐桌，这里简直是客人们的私人避风港。在这里欣赏墨尔本CBD宏伟景色或者在可以容纳8人的阳台作SPA赶走烦恼，与此同时还可以遥望雅普街上充满艺术、时尚和文化气息的场景。

1. Lift
2. Entrance
3. Suite
4. Deck
5. Floor lobby

1. 电梯
2. 入口
3. 客房
4. 接待台
5. 大堂

1. The entrance to the hotel
2. The dining area in John Olsen Penthouse
3. The bath on the terrace in John Olsen Penthouse
4. The pool inside
5. The lobby with Olsen's work
6. Deluxe Studio Suite
7. Spa Studio Suite
8. The lounge in Penthouse
9. The details in guestroom
10. Studio Suite Balcony
11. Two-Bedroom Open Plan Suite
12. Two-Bedroom Deluxe Balcony Suite
13. The bathroom in Two-Bedroom Open Plan Suite
14. The kitchenette in Two-Bedroom Open Plan Suite

1. 酒店入口
2. 约翰·奥尔森阁楼的就餐区
3. 约翰·奥尔森阁楼的露台浴池
4. 室内游泳池
5. 装饰着奥尔森作品的大堂
6. 豪华画室套房
7. 水疗画室套房
8. 阁楼套房休息区
9. 卧室的细节
10. 画室阳台套房
11. 开放式双卧室套房
12. 豪华阳台双卧室套房
13. 开放式双卧室套房的浴室
14. 开放式双卧室套房的小厨房

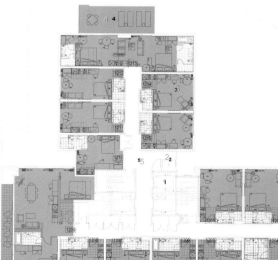

# English Art Hotel in New York

在纽约的英式艺术酒店

The Crosby Street Hotel is a hotel of Firmdale Hotel Group. Kit Kemp, in particular, the interior designer of the Crosby Street Hotel, retains the ownership as well. In 2008, Firmdale Hotel Group won the Andrew Martin International Interior Design Award and at present, its two representative hotels in London and New York are respectively remarked of first class level in artistic creation, which is largely attributed to the non-mainstream interior design style highly esteemed by Kit Kemp. Deeply influenced by sculptures, Kit Kemp expertly employs fabrics and artworks to make hotels abundant and charming. The Crosby Street Hotel, the first hotel she designed in USA, also introduces an environment of unexpected flair, warmth, and comfort. Works of British and Irish artists favoured by Kit Kemp such as Justine Smith, Jack Milroy, Anselm Kiefer, Callum Innes and Jaume Plensa are exhibited at the Crosby Street Hotel and of which a 10-foot-high stainless steel face, displayed in the lobby, is the most striking.

In addition, the Crosby Street Hotel is the centre for art shows. Many movies are premiered in the underground screening room, including James Franco's Howl. At the Hotel's opening ceremony, Jean Roman Seyfried, sculptor and photographer, made a time-lapse film showing the history of the Crosby Street Hotel transformed from a parking lot.

克罗斯比街酒店是弗姆戴尔酒店集团的管理酒店之一。特别的是，这家酒店集团的室内设计师和拥有者是同一人，她就是姬特·坎普（Kit Kemp）。2008年，佛姆戴尔酒店集团的室内设计获得安德鲁·马丁国际室内设计大奖，目前，在酒店集团的两个战略地——伦敦和纽约，佛姆戴尔被称具有极高的艺术水准，这是和姬特·坎普崇尚艺术的非主流的室内设计有着很大的联系。坎普的酒店受到绘画和雕刻的影响，她善于运用织物和艺术品使酒店丰满而充满神韵。克罗斯比街酒店是她设计开办的第一家在美国的酒店，这次同样将她温暖的英式风格和英式艺术观带入到酒店的设计之中。这里到处展示着坎普喜爱的英国和爱尔兰艺术家的作品，这些艺术家包括贾斯汀·史密斯（Justine Smith）、杰克·米尔洛依（Jack Milroy）、安塞姆·基弗（Anselm Kiefer）、卡勒姆·英尼斯（Callum Innes）和豪梅·普朗萨（Jaume Plensa）。最吸引人眼球的是大堂内摆放的一件高10英尺的不锈钢人头像。

除此之外，酒店还是展示艺术的集散地，在酒店地下的放映厅，许多电影在这里举办首映，其中包括詹姆斯·弗兰科主演的《嚎叫》。在酒店的开业庆典上，雕塑家、摄影家简·罗曼·西耶弗里德（Jean Roman Seyfried）为酒店制作了一部缩时（time-lapse）电影，描绘了克罗斯比街从一个停车场走到今天的变迁史。

# The Crosby Street Hotel

克罗斯比街酒店

**Completion date:** 2010
**Location:** New York, USA
**Designer:** Paul Taylor, AIA, President
**Size:** 7897 m$^2$
**Photographer:** Simon Brown

完成时间：2010年
项目地点：美国，纽约
设计师：保罗·泰勒（美国建筑师协会主席）
规模：7897平方米
摄影师：西蒙·布朗

The façade is made up of brick and limestone. The interior materials are varied but luxurious in every instance. German-oak floors are used throughout the building.

Located in the heart of SoHo, The Crosby Street Hotel is the first hotel outside of London for the premier luxury London hotelier, Firmdale Hotels. The SoHo property is built from the ground up: an 7897 m$^2$, 11-storey building with 86 guest suites.

Designed as the quintessential SoHo loft, the hotel contains signature features of the hotelier, including five and six fixture baths, graciously proportioned guestrooms, a 99-seat screening room, meeting rooms, a guest library, drawing rooms and a lounge, as well as a lobby bar and restaurant.

The hotel is filing for LEED Gold Certification, and will be one of the first New York City hotels to receive this designation. The property opened in September 2009. The interior is designed by Kit Kemp.

The hotel is filled with one-of-a-kind furnishings. Antiques and unique worldly materials combine to create an environment of unexpected flair, warmth, and comfort. The artwork and sculptures were all individually selected by the Interior Designer, Kit Kemp. Some of the pieces are from her personal collection.

The colour scheme is as eclectic as the materials used. The variation of colours, patterns, and textures all culminate to create a sense of freshness and fun from an unexpectedly inviting and new perspective.

建筑外立面材料为砖和石灰岩。内部材料多样，但每处都尽显奢华。建筑地面统一为德国橡木地板。

位于索霍区（美国纽约曼哈顿南部一地区）中心的克罗斯比街酒店是伦敦奢华型酒店集团Firmdale Hotels在伦敦外的首家酒店。该酒店占地面积7897平方米，分11层楼共86间客房。

作为索霍区阁楼设计的典范，该酒店拥有Firmdale Hotels的典型特征：5到6个固定浴池、优雅相称的客房、1间99座位的放映室、1间住客图书馆、多间会议室、画室以及大堂酒吧和休息室。

克罗斯比街酒店正申请LEED金牌认证，它将成为纽约市第一家获此称号的酒店。该酒店于2009年开业。内部由姬特·坎普设计。

酒店内家具独特，材质世所罕见，采用古风，营造出意想不到的温馨舒适的氛围。店内艺术品和雕塑由室内设计师基特·坎普亲选，部分来自她的个人收藏。

酒店配色方案与选料协调，多变的颜色、样式和质地集中营造一种全新的、令人意想不到的新奇感。

1. Bar in the hotel
2.4. Garden
3.5. Screening room
6. Drawing Room
7. Junior Suite
8. The bedroom in One & Two Suite
9. The layout of lobby
10. Artistic interior in Suite

1. 酒店内的酒吧
2、4. 花园
3、5. 放映室
6. 绘画客房
7. 标准客房
8. One & Two 套房的卧室
9. 大堂布局
10. 套房内充满艺术气息的内饰

| | |
|---|---|
| 1. Lobby | 1. 大堂 |
| 2. Reception | 2. 接待处 |
| 3. Drawing Room | 3. 绘画客房 |
| 4. Office | 4. 办公室 |
| 5. Pantry | 5. 食品储藏室 |
| 6. Restaurant | 6. 餐厅 |
| 7. Courtyard | 7. 庭院 |

# Mobile Art

移动的艺术

citizenM is a new Dutch hotel group that opened their first two hotels at Schiphol Airport and in Amsterdam City in 2008 and 2009 and citizenM Glasgow in 2010. citizenM Bankside will be the fourth hotel to open, and offers mobile citizens of the world affordable luxury in the heart of the city. The concept of the hotel is to cut out all hidden costs and remove all unnecessary items, in order to provide its guests with a luxury feel for a budget price.The design is focussed on citizenM's belief that a great bed and a simple and clean bathroom is all we need during a city or business trip. Concrete created the concept of citizenM as a holistic plan. It sets the boundaries for every creative process in all disciplines involved in the citizenM hotels. Concrete itself is responsible for the interior as well as the architectural design, with the corporation of local architects in the execution design, which not only converges diverse design ideas, but also infuses art elements into hotel design.

Inside the hotel, all is design for those travellers, highlighting the theme of mobile life. What the most distinguishing are the art elements putting into the kind of life. Mobile life creates mobile art. The hotel has 192 rooms of 14 sq m, all prefabricated in a factory and easy to transport, which is just reflecting the meaning of hotel's name. The signature of hotel interior design are the cabinet walls, spreading the seeds of art throughout the hotel and letting the art begin to move. Besides, by commissioning the huge artwork by Mark Titchner, citizenM wants to contribute to its local cultural environment. The artwork expresses the open and positive way of living and thinking of citizenM as well as its guests, the mobile citizen. Behind the courtyard, a full-length artwork forms the backdrop of the whole living room and canteenM area. The 25-meter-long vivid artwork created by the artist group AVAF ends the space dynamically with the arrival of art.

居民M是荷兰新兴的一家酒店集团，在2008年和2009年，最初的两家酒店分别在史基浦机场和阿姆斯特丹开业，随后在2010年格拉斯哥也开了一家居民M酒店。第四家酒店将在伦敦的河岸区开业，在这座城市的中心酒店将提供给旅行者世界级的奢华。酒店的理念是去除所有隐性成本，移除所有不必要的物件，让客人用低廉的价格就能得到奢华的感受。酒店的设计忠于居民M品牌的宗旨，舒适的大床和简单干净浴室，这是客人在一座城市旅行的所有必须。混凝土设计事务所将这个宗旨理念作为全盘的设计计划，贯彻至居民M酒店的每个设计环节之中。混凝土设计负责酒店的室内和建筑设计，并与当地建筑师一起落实了设计的完成，这不但汇聚了多方设计灵感，更让整个酒店的设计充满艺术气息。

在酒店室内，设施都依客人的旅行方便而定制，突出移动式生活的主题，然而最有特色的是在这种生活方式中加入的艺术元素，移动的生活也成就了移动的艺术。酒店共有192间14平方米的客房，它们都在一家工厂预先建造组合，运输至客房内正符合了酒店名字的寓意。酒店内的标志是休息区的橱柜墙，它将酒店的艺术感蔓延在整个酒店的公共区内，也让艺术真正的移动起来。此外，酒店委托艺术家马克·蒂奇纳尔设计了一幅巨大的艺术作品，以向当地的文化环境致敬。艺术品所表达出居民M开放和积极的生活以及思考方式，这同样也是酒店客人追求的一种方式：移动的居住方式。在庭院的后部，一排艺术品组成了整个客厅区和餐厅区的背景。排成25米长的生动艺术品由艺术集团AVAF创作，为整个空间的设计做出生动富有艺术感的结尾。

# citizenM Hotel Bankside London, UK

伦敦河岸区居民M酒店

**Completion date:** 2012
**Location:** London, UK
**Designer:** Concrete
**Size:** 192 rooms
**Photographer:** Richard Power for Concrete

完成时间：2012年
项目地点：英国，伦敦
设计师：混凝土设计事务所
规模：192间客房
摄影师：理查德·鲍威尔

The building is located on the Southbank of the Thames in the Southwark area. Positioned on the corner of Lavington street and Southwark street it is only a few steps south from the Tate Modern. Southwark is an upcoming area with an interesting mix of functions including art galleries, boutique shops and restaurants. The building itself consists of two 6-storey blocks divided by an internal courtyard. The volume, as well as the architecture seeks a relationship with the former industrial area by its robust form and materialisation.

Sand coloured stone panels with special designed brickwork pattern are referring to the warehouse architecture of the neighbourhood. Within this robust volume big glass windows of the rooms are pushed out.

The ground floor is one open plan. Diverse areas are divided and made intimate by open cabinets, bespoke furniture pieces and the fully glazed courtyard. The building has two entrances, one entering the check-in area, the other one leading towards canteenM, the café and F&B area. The wooden entrance box gives a warm welcome. The hotel does not have a service desk but a table with six self-check-in terminals. One of citizenM's ambassadors is present to guide you through the check-in procedure (30 seconds). A full length cabinet next to the check in table guides the guest towards the elevator. The cabinet is styled with items and books, as well as some great pieces of art, all to inspire the people. A small shop with cool magazines, newspapers, citizen goods and all necessary items that you need but tent to forget to bring with you is housed in the cabinets as well. Part of the cabinet wall is a living room area with a large fireplace. Opposite the elevators, next to the glazed courtyard, a wooden iconic helical stairs invites you to enter the societyM area on the first floor. Opposite the fireplace a blown up book houses the shop in shop of Mendo Books. The latest and most interesting books on art, design, photography and other interesting social cultural items are there to browse through and for sale.

The rooms are designed in cooperation with Vitra, using their furniture collection. Between two cabinet walls a large living area under a cloud of Modernica lamps is located right behind the full storey glass windows facing the street. Six small dining, drinking and working tables pop out of the homey styled cabinets. Centrepiece next to the bar is the lounge area: a large leather poof is positioned on a mixed up union jack carpet. Two 'bookcase' benches flank the lounge area, filled with interesting and inspiring books to read and flip trough.

The courtyard is positioned right beside the living rooms and canteen area. Large glazed windows bring daylight deep into the building. The wooden terrace decking is a continuation of the interiors' wooden floors. Silver birch trees, one placed in a red wooden tree-bench, create an intimate outdoor area. Red coloured terrace seating provide place to hang around. Lanterns are hung high up in the yard for a great evening feel.

The iconic wooden stairs opposite the elevators brings you to societyM on the first floor, citizenM's business area. It is created for a new type of worker: those who aren't bound by offices or office conventions, wanderers doing business wherever the connectivity's good and the coffee fresh. We call these folk: business nomads.

Seven creating rooms are positioned around the courtyard. All rooms fully equipped

with audiovisual equipment, white and black walls for notes, theories and scribbles, and the bespoke cabinet filled with inspirational items and books. The fully glazed corridor alongside the courtyard, in front of the creating spaces, houses some low seating facilities to withdraw from the meeting or make a phone call.

All the citizenM guestrooms are constructed off site. This means the outside measurements are restricted to enable the transport to the building-site. Within these measurements the guestroom will be continually evaluated and improved for every hotel. The bed stands in front of a floor-to-ceiling, wall-to-wall window and is super-king size, 2,2 x 2,0 m. Its white bed-linen and pillows offer a lounge area to watch TV on the flat LCD screen, together with the stuffed animal named Marvin (designed by Gewoon) that's placed in every room. The Night tables on both sides of the bed contain plisse-shaded lamps. One of them is designed in a such way that it also can be used as a desk. Underneath the bed is a huge drawer to store an opened suitcase or other personal belongings. Sockets next to the bed-front make it possible to connect your laptop or telephone. To create maximal space, the bathroom elements are placed separately in the room. The wet-room, which contains rain- and hand-shower and the toilet, is a half-rounded cabin and is situated on one side of the room. The wet-room is constructed from frosted glass to create transparency yet ensure the feel of privacy. The translucent ceiling lights up in any colour you want and become a signature element in the room. On the opposite side of the wet-room there is a vanity made of Corian. The desk contains the washbasin, make-up mirror with lighting and storage space. A minibar is also placed into the vanity. To create a theatrical atmosphere the room contains diverse light-sources. Among them a large RGB LED-string above the translucent ceiling offers the opportunity to change colours of the room.

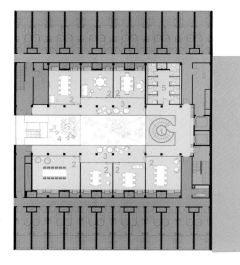

1. Helical stair to ground floor
2. Creating rooms
3. Corridor for informal meet, wait and call
4. Patio
5. Restrooms
6. Ironing room
7. Bedrooms
8. Back of house

1. 通往一层的螺旋楼梯
2. 创意空间
3. 非正式会议室走廊
4. 天井
5. 洗手间
6. 熨衣间
7. 客房
8. 后场

酒店的建筑坐落在萨瑟克区泰晤士河的南岸。位于拉文顿街和萨瑟克街的夹角处，往南距离泰特现代美术馆只有几步之遥。萨瑟克是一个新兴的区域，这是一个混合了不同功能建筑的有趣区域，包括美术馆、精品商店和餐馆。建筑本身由两座6层的建筑组成，中间由一个内部庭院分开。从建筑的体积和风格，可以看出设计者通过充满活力的建筑形式和物化手段在寻找与先前工业区之间的某种关联。

酒店的第一层是一个开放的空间。不同的区域被分开，但开放的酒柜、定制的家具和玻璃墙围出的庭院让空间显得相互呼应。建筑有两个入口，一个通往入住区，另一个则通往M餐厅、咖啡厅和餐饮区。盒子状的木质入口营造好客的氛围。酒店不设有签到服务台，但是有6个自动入住签到终端系统。酒店的接待人员会引导客人在30秒内完成入住程序的办理。登记台旁边的长橱柜引导着客人通往电梯间。橱柜内摆着各种艺术装饰品和书籍，激发着人们的灵感。摆着潮流杂志、报纸、日常用品和旅行必备物件的橱柜组成了一个小商店，客人可以找到一切容易忘记带上路的旅行所需。橱柜的一部分组成了休息区，还设有一个壁炉。背对电梯，挨着庭院，一段螺旋状的楼梯引导客人通向二楼的M社会会议空间。在壁炉的对面，一叠叠书组成了Mendo书店。这里有最新、最有趣的艺术类、设计类、摄影类和其他相关社会文化的书籍可供客人阅读和购买。

1. The view from guestroom to the courtyard
2. The living room on ground floor
3. canteenM
4. Mendo bookshop next to helical stairs
5. The helical stairs to first floor
6. societyM
7. The cabinets with books in Mendo bookshop

1. 从客房看庭院
2. 一层的起居空间
3. M餐厅
4. 挨着螺旋楼梯的Mendo 书店
5. 通往二层螺旋楼梯
6. M社会
7. Mendo书店填满书的橱柜

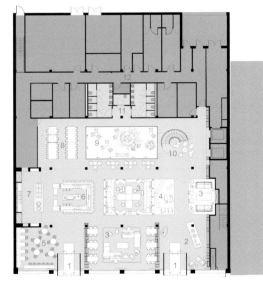

1. Entrance citizenM
2. Check-in area
3. Living room
4. Mendo bookshop
5. Coffee corner
6. Bar
7. Kitchen and buffet
8. Working area
9. Patio
10. Helical stair to creating spaces
11. Restrooms
12. Back of house

1. 酒店入口
2. 签到区
3. 起居空间
4. Mendo书店
5. 咖啡厅
6. 酒吧
7. 厨房和自助餐区
8. 工作区
9. 露台
10. 通往创意空间的螺旋楼梯
11. 洗手间
12. 酒店后场

8-9. The cabinets filled with artworks
10. Coffee corner in living room
11. Working area in living room
12. The vanity in guestrooms
13. The guestroom with wet-room

8、9. 填满艺术品的橱柜
10. 起居空间的咖啡厅
11. 起居空间的工作空间
12. 客房浴柜
13. 客房和客房淋浴间

设计师与维特拉公司合作，在起居空间配置了他们的精选家具。在两面橱柜墙之间是一个很大的起居空间，透过临街的整层玻璃窗墙可见一片摩登妮卡灯饰安装在起居空间。6个小型餐饮区和工作桌不规则的点缀在舒适、个性的橱柜之间。靠近酒吧中心的是酒廊区：大皮革沙发混搭带有英国国旗图案的地毯。两个"书箱"长椅摆放在酒廊区的两侧，里面填满了有趣且充满启发性的书籍以供客人阅读和品味。

庭院就设置在客厅区域和餐饮区旁。巨大明亮的玻璃将日光深层地引入建筑之中。木制露台是室内木地板的一种延伸。这里有许多银桦树，其中一些被种在红色的木质长凳之中，营造出温馨的室外空间。红色的露台座椅吸引着客人到此处休息、消遣。一些灯笼挂在庭院的半空，在夜晚别有一番风韵。

时尚的木质楼梯背对电梯间，带客人来到二层的商务区域M社会。这里是为新兴工作者创造的空间：那些不被禁锢在办公室和会议室的工作者，他们选择在有畅通的网络和新鲜咖啡的环境中工作。这些人被称为：商务游牧民。7个创意空间被设置在庭院的四周。所有的房间都配备了视听设备，白色和黑色的墙供他们作记录、研究学问和随意涂写，定制的柜子被艺术品和书籍填满。在这些创意空间前，庭院旁明亮的走廊摆设了一些座椅，供闭会后的客人休息或者打电话使用。

所有的居民M客房是在建筑外组建的。这意味着外部的测量值是固定的，这样可以保障运输的畅通。在这些测量值的限制下，客房将被不断的测量，根据每家酒店的情况不断进行改进。床设置在落地窗前，是2.2米x 2.0米的大床。白色的床单和枕头提供了一个休闲区域供观看挂在墙上的LCD电视，此外在每间客房的床上都有一个名叫马尔文的填充动物玩具（由格温设计）。床头柜在床的两侧，上面有带泡泡纱灯罩的台灯。其中一个床头柜还可以当做书桌使用。在床下是一个巨大的抽屉，以储备旅行箱和其他个人财产。床边的插座可以让客人方便使用笔记本和电话。为了制造出最大的空间，浴室的元素被分设在房间内。淋浴间包含花洒、淋浴喷头和马桶，浴柜则在客房的另一侧。淋浴间由磨砂玻璃围住，营造出透明感，确保客人隐私。半透明的天花板可以按客人想要的颜色亮起来，这成为了客房的标志特色。在淋浴间的另一侧是可丽耐浴柜。桌面安有洗手盆，带灯的化妆镜和储藏空间。浴柜的一部分还可以用来当做迷你酒吧台使用。为了营造戏剧般的氛围，客房内装设不同种类的光源。一个大的三原色LED灯带安装在半透明的天花板内，提供了改变客房内颜色的机会。

# A Journey to Diverse Arts Neo-Baroque

新巴洛克的多元艺术之旅

Since it was opened, Byblos Art Hotel Villa had been keeping the easy and vigorous ambience, and enjoying the happiness of life as its brand mark. It is good at using lively colour, geometry, embroidery totem to reveal the perfect combination of the fashion and modern art. Byblos Art Hotel Villa is the masterwork of Dino Facchini who is the father of Byblos. Travel and modern art are combined together in this hotel. In 2000, Dino Facchini bought this manor which was originally built in 16th century. He invited the famous designer Alessandro Mendin to renovate the hotel, which can not only highlight its classical details but also express its beauty of modern art. Here exhibits many works of the world famous artists such as Vanessa Beecroft, Anish Kapoor, Peter Halley, Sol Lewitt, and Gio Pomodoro. Alessandro Mendin decorated the villa with futuristic furniture designed by him and the modern art works privately collected by Dino Facchini. Each room is filled with unique designs for ponder, which can express the colourful trend and culture such as exotic landscapes, modernism and internationalism, in the meantime perfectly combined with modern new materials and techniques. Byblos Art Hotel Villa is a natural combination of classicism charm, modern spirit and the style of the Byblos brand. It takes visitors back to the glorious days of this grand building, and brings them a kind of experience of Neo-Baroque country life.

比布鲁斯品牌的设计以颜色鲜艳活泼、几何图案大胆果断、刺绣精美绝伦著称，堪称时尚和现代艺术的完美结合。自品牌创立以来，比布鲁斯一直坚持轻松惬意、活力充沛、享受生活的品牌精神。比布鲁斯艺术别墅酒店是毕伯劳斯之父迪诺·法琦尼（Dino Facchini）的杰作。这家酒店把旅游和当代艺术结合在了一起。迪诺·法琦尼在 2000 年买下了这座始建于 16 世纪的庄园，请室内著名设计师亚历山德罗·门迪尼（Alessandro Mendin）进行翻新，即突出它的古典细节又能展现现代艺术之美。这里展出多件世界著名的艺术家，如 Vanessa Beecroft、Anish Kapoor、Peter Halley、Sol Lewitt、Gio Pomodoro 等的作品。亚历山德罗·门迪尼将其设计的未来主义派家具加上迪诺·法琦尼私人收藏的当代艺术品对别墅进行装饰，让每间房间都布满可供玩味的独特设计，表现出多姿多彩的潮流和文化，例如异域风情、现代主义和国际主义，同时完美的结合当代的新材料和技术。比布鲁斯艺术别墅酒店是古典主义神韵、现代精神以及比布鲁斯品牌的风格的自然融合，它将游客们带回到这座宏伟建筑往昔的辉煌岁月，给他们带来一种新巴洛克式的乡村生活体验。

# Byblos Art Hotel Villa Amistà

比布鲁斯艺术别墅酒店

**Completion date:** 2005
**Location:** Verona, Italy
**Designer:** Alessandro Mendini, Gianfranco
**Size:** 5000 m²
**Photographer:** Tony Cragg

完成时间：2005年
项目地点：意大利，维罗纳
设计师：亚历山德罗·门迪尼，詹弗兰科·帕格拉
规模：5000平方米
摄影师：托尼·克拉格

The large building featuring a 15th century façade by architect Michele Sanmicheli (1489-1559) had been abandoned for some time before being carefully renovated thanks to Dino Facchini's and his family's passion for art. The owners of the prestigious Byblos fashion label, have made accurate renovations both in terms of the architecture and the paintings and décor. Restoration has been carried out on the central and more monumental part of the villa and also on the two towers, the small church dedicated to Saint Rocco and the farm buildings. The complex renovation project was completely designed by Alessandro Mendini and his Atelier. The extensive garden, covering an area of twenty thousand square metres, was designed by Gianfranco Paghera and offers an evocative setting, an "earthly paradise" with fountains and swimming pool, where guests can enjoy a unique atmosphere by smell, sight and touch all the year round.

A precise combination of these different visual elements results into a fascinating set design, an accurate scenic and museum-like setting to enable an outstanding hospitality where the glamorous prét à porter fashion label Byblos is melted into a synergy with the splendour of an ancient past and a modern spirit. The aim was to bring this magnificent summer villa back to its original splendour and take the hotel guests into a Neo-Baroque rural experience among past, present and future. Furthermore, some of the most impressive furniture items and objects (such as chairs, pillows, fabrics, silverware, candelabra, plates, cups and others) have been exclusive designed for the hotel by the well-known architect and designer Alessandro Mendini and, belonging to the BYBLOS CASA collection, can also be purchased by the guests.

The collection represents streams such as ethnic style, revival or international design lines and combines contemporary technological materials. Byblos Casa aims for a high-level clientele, although it offers also more accessible, less expensive products.

这座由建筑师米歇尔·森密切里（1489-1559）设计的15世纪大型建筑在被精心的翻修之前曾一度要被废弃。由于迪诺·法琦尼家族及其他他们对艺术的热爱，酒店被精心翻修。比布鲁斯的拥有者将这个有威望的"时尚标签"在建筑、室内陈设和装饰上进行了精确的翻新。酒店的中心部分、别墅的大部分以及两座塔楼、圣罗克小教堂和农场建筑都被进行了修复。整个翻新项目是由设计师亚历山德罗·门迪尼和他的工作室负责整体设计的。詹弗兰科·帕格拉设计并扩建了一个面积为20,000平方米的花园，同时还提供了一个能够唤起回忆的场景，一个由喷泉和泳池组成的"人间天堂"，一年四季客人们在这里都可以通过嗅觉、视觉、触觉体验到一种独特的氛围。

这些不同的视觉元素的精确结合产生出的是一种令人着迷的场景设计，一个名副其实的风景胜地以及拥有出色接待条件的博物馆式场景，极富魅力的高端时尚标签——比布鲁斯被融入到一种由古代和现代精神相结合产生的辉煌之中。翻修工作的目标是将这个宏伟的夏季别墅恢复到它最原始的辉煌时期，并且给客人带来一种新巴洛克的乡村生活体验。此外，最让人印象深刻的一些家具和物品（例如椅子、枕头、织物、银器、枝状烛台、盘子、杯子等）都是由知名的建筑师和设计师亚历山德罗·门迪尼特别设计，属于比布鲁斯酒店家居用品系列，客人们可以在酒店购买到。

家居用品系列表现出的是多种思想的源泉，例如异域风情、复兴精神和国际设计，同时这些用品结合了先进的技术材料。尽管比布鲁斯酒店的目标是为高阶层的顾客服务，但酒店同样提供更多方便使用、价格低廉的产品。

1. The exterior of hotel
2. The outside pool
3. The interior of Beatriz Room
4. Anna Grilli minibar
5. Giardino d'Inverno with a relaxing atmosphere
6. Dining table with a work by Vanessa Beecroft
7. A work by Vanessa Beecroft
8-9. ESPACE
10-12. The bedroom in Executive Junior Suite Yellow
13. Executive Suite
14. A corner in guestroom

1. 酒店外立面
2. 户外游泳池
3. 贝特丽姿餐厅内饰
4. 安娜吉利迷你酒吧
5. 充满轻松氛围的冬季花园餐厅
6. 用凡妮莎贝克福特作品装饰的餐桌
7. 凡妮莎贝克福特的艺术作品
8、9. ESPACE 水疗中心
10-12. 行政标准黄色套房的卧室
13. 行政套房
14. 客房内一角

# European Art Geco Gem with Oriental Feelings

充满东方风情的欧洲装饰艺术瑰宝

Prague Art Deco Imperial Hotel, which is a European Art Deco gem with oriental feelings, was built in 1913-14 as a luxury hotel. Its façade is of typical geometry Art Deco style combined with some Cubism design elements, which makes it massive and dignified, simple and natural. However, such an exterior is easy to make people have a wrong association with its interior decoration. At the moment that you step into the hotel, you'll be sure to marvel at its splendid and magnificent interior design which is also combined with many kinds of cultures and art styles. Ancient Egyptian, Arabic, European and modern art styles are skillfully combined together just like that it is made by nature. The entry hall of the hotel is spacious with colourful brick walls, Mosaic ceilings, plant relief and figure relief of Egyptian and Mediterranean styles with primitive surface decorations. Just facing the entry hall, the marble stairs gracefully spiral to the first floor, which makes the hall more spacious and magnificent. The lobby, whose whole tone is beige, is decorated with textile furnishings of concise and decent modelling, which gives a comfortable feeling. The most prestigious in the hotel is its royal café and restaurant. Its hall is grand and as gorgeous as the European royal palace. The whole ceiling is tiled with late Art Nouveau style Mosaic which is very exquisite. The form of the pillars and beams is similar to the ancient Egyptian architecture style, which is solemn and sacred. All the walls and pillars are decorated with awesome relief decorative brick. Complicated and exquisite patterns of flowers, animals and characters give a strong mysterious oriental feeling. It can be said that Prague Art Deco Imperial Hotel is a work of art in itself.

布拉格皇家装饰艺术酒店始建于1913–1914年，定位为豪华酒店。外立面是典型的几何装饰主义风格，并融入了一些立体主义设计元素，厚重端庄、朴素自然。但如此外观也很容易使人对其室内装饰风格产生错误的联想。当你初次踏进酒店的那一刻一定会被它富丽堂皇、美轮美奂，并且融合了多种文化和艺术风格的室内装饰折服。古埃及、阿拉伯、欧洲以及现代的艺术风格被巧妙的组合在一起，浑然天成。酒店门厅宽敞气派，彩色铺砖墙面、马赛克天花板，表面装饰古朴的埃及、地中海风格的植物、人物浮雕。正对门厅，大理石阶梯优雅的盘旋通往二楼，衬托得大堂越发宏伟大气。大堂整体呈现米黄色色调，布置着造型简洁大方的织物家具，给人舒适温馨的感觉。酒店最富盛名的是皇家咖啡厅与餐厅。大厅宏伟壮观，犹如欧洲皇家宫殿一般华美。用色主要为白色和金色，白色给人以纯洁高雅的感觉，金色则散发尊贵、奢华的气质。天花板全部是用后新艺术派风格马赛克拼成，非常精美。柱梁形式类似古埃及建筑风格，庄重、神圣。所有的墙面、柱面都装饰着令人叹为观止的浮雕装饰砖。繁复精美的花卉、动物和人物图案透出神秘的浓浓东方风情。可以说，布拉格皇家装饰艺术酒店本身就是一件艺术品。

# Prague Art Deco Imperial Hotel
## 布拉格皇家装饰艺术酒店

**Completion/Latest renovation date:** 1914/2007
**Location:** Prague City, Czekh
**Designer:** Jan Melka, Jaroslav Benedikt, Josef Drahoňovský, Ivo Nahálka
**Size:** 86 Deluxe rooms, 34 Executive rooms, 5 Junior Suites, 1 Imperial Suite
**Photographer:** Prague Art Deco Imperial Hotel

**完成/翻新时间：** 1914年/2007年
**项目地点：** 捷克，布拉格
**设计师：** 简·梅尔卡，亚罗斯拉夫·内迪科特，约瑟夫·德拉候纽夫斯基，伊沃·那哈尔卡
**规模：** 86间豪华客房，34间行政客房，5间普通套房，1间皇家套房
**摄影师：** 由布拉格皇家装饰艺术酒店提供

Prague Art Deco Imperial Hotel was built during 1913-14 as a luxury hotel and its geometric Art Deco exterior with Cubism components conceals an interior embellished by precious late-Art Nouveau mosaic. The extraordinary entrance hall with its colourfully tiled walls and mosaic ceiling is garnished by floral and figural decorations inspired by Egyptian and Mediterranean cultures, while the grand marble staircase complements the imposing space of the lobby. The magnificent rooms of the Café and Restaurant Imperial boast exquisite ceramic mosaics, and the superb tile-decoration of the walls and pillars overflow with rich floral and animal ornamentation recalling oriental and Moorish arts.

Since its opening, the hotel has been recognised for its excellent service and hospitality, and the Café Imperial has long been rated as one of the most sought-after places in Prague. As a result of its outstanding art & architectural value, the Prague Art Deco Imperial Hotel was classed amongst the city's historically listed monuments and thanks to the extensive reconstruction, with special attention to the historical details, this treasure of turn-of-the-century architecture has been restored to its former splendour and is ready to satisfy even the most sophisticated travellers.

The first written reference to the house standing on the corner of Na Porici and Zlatnicka streets dates back to 1383, when the existing building was joined with its neighbouring construction into one single complex. At the end of the thirty-year war during the Swedish siege of Prague the house burned down, but later, thanks to the extensive reconstruction work around 1730, it was turned into an Inn entitled 'The Black Eagle'. The yard wings were enlarged during building renovations carried out in 1840 subsequent to which it became well-known as the 'Hotel At the Black Eagle'. One of the prominent owners of the hotel was Barbara Serafinova, who generously sponsored the charity activities of her sister Anna Naprstkova, and as a part of her inheritance the Hotel At The Black Eagle became the property of the Naprstek family. This provided the hotel with its most famous owner in 19th century - Vojtech Naprstek - the famous Czech writer, politician and patron who devoted much of his time and experiences to travelling and collecting valuable items worldwide. Later in the 19th century the hotel came under ownership of the Czech Industrial Museum Foundation, established by Vojtech Naprstek, but was later demolished as a part of Prague' general urban renewal in 1913. During 1913-1914 the architect Jaroslav Benedikt designed and built the luxury Art Deco Hotel Imperial, the magnificent ceramic interiors being designed by Jan Benes complemented by plastics from Josef Drahonovsky. From 1948, when the Trade Unions' Association had taken over its management, the hotel provided preferential accommodation for the guests and members of the trade unions, until its operation as a hotel was disrupted in the1980s. In 2005-2007 extensive reconstruction work was undertaken to restore this precious and internationally-recognized unique gem of 20th century architecture to its former glory, re-establishing the Prague Imperial's reputation as the most outstanding and spectacular luxury hotel in the city.

布拉格皇家装饰艺术酒店始建于1913-1914年，这家豪华酒店有着混合了装饰艺术及立体派艺术的建筑外表，和装饰着珍贵晚期新艺术风格的马赛克的内部设计。在非同寻常的入口大厅，墙壁上贴着五颜六色的瓷砖，天花板上装饰着印有花朵和人物形象的马赛克，这些图案的灵感来自于埃及和地中海文化。与此同时，壮观的大理石阶梯使大堂的空间显得更加气势宏伟。在皇家咖啡厅与餐厅，富丽堂皇的空间内装饰着陶瓷马赛克和工艺卓绝的陶瓷装饰品，立柱的周身装饰着丰富的花朵及动物装饰图案，唤起了客人对东方及摩尔艺术的遐想。

自从酒店开业之日起，酒店就因其卓越的服务和食宿条件受到各方认可。皇家咖啡厅被评为布拉格最值得一去的景点之一。凭借其非凡的艺术与建筑价值，和扩建性的重建工程，布拉格皇家装饰艺术酒店位列城市历史纪念建筑的名单之中，特别注重对于历史细节的保护，这座世纪交替时建成的珍宝级建筑被修复出当年的辉煌景象，并已经做好满足品位最高游客们的准备。

第一次关于这家位于那波里奇街和紫拉妮卡街交汇处的酒店的文献记载要追溯到1383年，在当时这座建筑与其他邻近的建筑组成了一个独立的建筑综合体。在上世纪30年代末，瑞典人占领布拉格期间，这座建筑被烧毁，但在之后1730年建筑被整修扩建，改造成了"黑鹰"旅馆。在1840年，建筑的庭院两侧被扩建，紧接着著名的"黑鹰"酒店成立了。酒店著名的拥有者之一是芭芭拉·萨拉夫诺娃，她资助着姐姐安娜·纳普勒斯科娃的慈善活动，作为她的遗产的一部分，黑鹰酒店成为了纳普勒斯科家族的财产。这使得酒店在19世纪有了一位最著名的业主——沃伊捷赫·纳普勒斯科，捷克著名的作家、政治家和资助人，他热爱旅行和收藏，他用大部分的时间环游世界并收藏有价值的艺术品。在19世纪，酒店落到了捷克工业博物馆基金会的名下，这家基金会由沃伊捷赫·纳普勒斯科建立，但是不久之后的1913年由于布拉格整体城市的重建工作再次被拆毁。在1913年-1914年期间，建筑师亚罗斯拉夫·内迪科特设计并建造了这座奢华的皇家装饰艺术酒店，宏伟的陶瓷内饰由让·贝奈斯设计，酒店外部的修补由约瑟夫·德拉候迪夫斯基负责完成。1948年，酒店接由贸易联盟协会掌管，开始为贸易联盟的客人和成员提供优质的住宿环境，直到20世纪80年代，这座建筑不再作为酒店使用。在2005年-2007年期间，酒店再次扩建翻修，这次翻修使这座珍贵、享有国际知名度的20世纪珍宝建筑恢复了之前的光辉，重建了布拉格皇家的威名，也让其成为这座城市中最耀眼、最奢华的酒店。

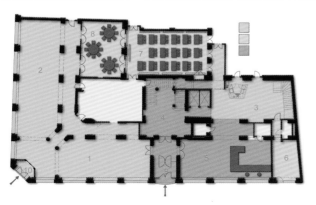

| | | | |
|---|---|---|---|
| 1. Café Imperial | 1. 皇家咖啡厅 | 1. Imperial Diamond ballroom | 1. 皇家钻石宴会厅 |
| 2. Restaurant Imperial | 2. 皇家餐厅 | 2. Imperial Café | 2. 黄金咖啡厅 |
| 3. Prime Bar | 3. 繁荣酒吧 | 3. Imperial cocktail bar | 3. 皇家鸡尾酒吧 |
| 4. Lobby | 4. 大厅 | 4. A corner in the bedroom of Executive room | 4. 行政客房一角 |
| 5. Reception | 5. 接待处 | 5. The bedroom in Imperial Suite | 5. 皇家客房卧室 |
| 6. Board room | 6. 董事会议室 | 6. The interior of Executive room | 6. 行政客房内饰 |
| 7. Diamond | 7. 钻石大会议室 | 7. The interior of Imperial Suite | 7. 皇家客房内饰 |
| 8. Topas | 8. 托帕斯宴会厅 | 8. The details of marble bathroom | 8. 大理石浴室细节 |
| 9. Main entrance | 9. 主入口 | 9. The classic interior of lobby | 9. 经典的大堂内饰 |
| 10. Café entrance | 10. 咖啡厅入口 | | |

# Index 索引

1. Hotel Le Bellechasse
   Paris, France
   info@lebellechasse.com
   Christian Lacroix

2. Hotel du Petit Moulin
   Paris, France
   agorand@hlgrp.com
   Christian Lacroix Design/Bastie Architect

3. Hotel Notre Dame
   Paris, France
   gwendolinedemorge@gmail.com
   Christian Lacroix

4. Maison Moschino
   Milan, Italy
   martinabarberini@mobygest.it
   Moschino

5. La Maison Champs Elysees
   Paris, France
   severine@re-active.fr
   Maison Martin Margiela

7. Armani Hotel Dubai
   Dubai, UAE
   MBaldonadi@armanihotels.com
   Giorgio Armani

8. Hotel Missoni Edinburgh
   Edinburgh, UK
   stephanie.maclachlan@stripecom.co.uk
   Rosita Missoni, Matteo Thun & Partners

9. Palazzo Versace
   Main Beach, Gold Coast, Australia
   pr@palazzoversace.com
   Soheil Abedian, Gianni Versace

10. Casa Camper Barcelona
    Barcelona, Spain
    alefebvre@camper.es
    Fernando Amat from Vinçon, Jordi Tió

11. Hotel Lungarno
    Florence, Italy
    press@lungarnocollection.com
    Michele Bönan, Nino Solazzi

12. Portrait Suites
    Rome, Italy
    press@lungarnocollection.com
    Michele Bönan

13. Tcherassi Hotel + Spa
    Cartagena, Colombia
    lanny@em50.com
    Silvia Tcherassi

14. Claridge's
    London, UK
    akadic@maybourne.com
    David Collins, David Linley, Diane von Furstenberg, Thierry Despont, Dale Chihuly

15. Alma Schlosshotel im Grunewald
    Berlin, Germany
    reservations@schlosshotelberlin.com
    Karl Lagerfeld

16. The Beverly Hills Hotel and Bungalows
    Beverly Hill, USA
    jenna.duran@dorchestercollection.com
    Adam Tihany

17. UXUA Casa Hotel
    Trancoso, Brazil
    pr@uxua.com
    Wilbert Das

18. Tortuga Bay
    Punta Cana, Dominican Republic
    agorand@hlgrp.com
    Oscar de la Renta

19. The g Hotel
    Galway, Ireland
    jpasztor@theg.ie
    Philip Treacy

20. The May Fair Hotel, London
    London, UK
    bjohnson@mmgyglobal.com
    Michael Attenborough

21. W Paris - Opéra
    Paris, France
    Maud.Montabone@starwoodhotels.com
    W Global Brand Design & Rockwell Group
    Europe (RGe) – Director: Diego Gronda

22. Hotel BLOOM!
    Brussels, Belgium
    info@marionflipse.com
    Bronwynn Welch

23. Gladstone Hotel
    Toronto, Canada
    jeremy@gladstonehotel.com
    Christina Zeidler and 37 local artists

24. Casa do Conto, arts&residence
    Porto, Portugal
    info@pedraliquida.com
    Pedra Líquida

25. The Olsen Hotel
    Melbourne, Australia
    mint@themintpartners.com.au
    Chris Hayton

26. The Crosby Street Hotel
    New York, NY, USA
    info@stonehilltaylor.com
    Paul Taylor, AIA, President

27. citizenM Hotel Bankside London
    Landon, UK
    info@concreteamsterdam.nl
    Concrete

28. Byblos Art Hotel Villa Amistà
    Verona, Italy
    info@fix.com
    Alessandro Mendini, Gianfranco

29. Prague Art Deco Imperial Hotel
    Prague city, Czekh
    jan.visek@hotel-imperial.cz
    Jan Melka, Jaroslav Benedikt, Josef
    Drahoňovsk, Ivo Nahálka

# 图书在版编目（CIP）数据

时尚艺术酒店 /（澳）惠特克编；于芳，李红译. --
沈阳：辽宁科学技术出版社，2013.6
　ISBN 978-7-5381-7945-3

Ⅰ. ①时… Ⅱ. ①惠… ②于… ③李… Ⅲ. ①饭
店－建筑设计－作品集－世界 Ⅳ. ①TU247.4

中国版本图书馆CIP数据核字(2013)第054717号

出版发行：辽宁科学技术出版社
　　　　　（地址：沈阳市和平区十一纬路29号　邮编：110003）
印　刷　者：利丰雅高印刷（深圳）有限公司
经　销　者：各地新华书店
幅面尺寸：240mm×290mm
印　　张：40
插　　页：4
字　　数：80千字
印　　数：1～1500
出版时间：2013年 6 月第 1 版
印刷时间：2013年 6 月第 1 次印刷
责任编辑：陈慈良　于　芳
封面设计：曹　琳
版式设计：曹　琳
责任校对：周　文
书　　号：ISBN 978-7-5381-7945-3
定　　价：298.00元

联系电话：024-23284360
邮购热线：024-23284502
E-mail: lnkjc@126.com
http://www.lnkj.com.cn
本书网址：www.lnkj.cn/uri.sh/7945